OUR WILD FAMILIARS

OUR WILD FAMILIARS

How Animals Are Adapting to Cities
and Reshaping the Natural World

DAN WERB

CROWN
NEW YORK

CROWN
An imprint of the Crown Publishing Group
A division of Penguin Random House LLC
1745 Broadway
New York, NY 10019
crownpublishing.com
penguinrandomhouse.com

Art appearing on pages ii–iii: Adobe Stock / Alexander Potapov (squirrel); Adobe Stock / Arrows (tiger); Adobe Stock / dipu (eagle); Adobe Stock / Rizwan Ahmad (baboon); Adobe Stock / MDFAZAR (pigeon); Adobe Stock / Shahadat (raccoon)

Library of Congress Cataloging-in-Publication Data
Names: Werb, Dan author. Title: Our wild familiars : how animals are adapting to cities and reshaping the natural world / Dan Werb. Identifiers: LCCN 2025050913 (print) | LCCN 2025050914 (ebook) | ISBN 9780593799635 hardcover | ISBN 9781039056824 international edition | ISBN 9780593799642 ebook Subjects: LCSH: Urban ecology (Biology) | Urban animals—Adaptation |Nature—Effect of human beings on | Human-animal relationships Classification: LCC QH541.5.C6 (print) | LCC QH541.5.C6 (ebook) LC record available at https://lccn.loc.gov/2025050913
LC ebook record available at https://lccn.loc.gov/2025050914

Hardcover ISBN 978-0-593-79963-5
Ebook ISBN 978-0-593-79964-2

Editor: Kevin Doughten | Editorial assistant: Jessica Jean Scott | Production editor: Natalie Blachere | Text designer: Amani Shakrah | Production: Heather Williamson | Copy editor: Dianna Stirpe | Proofreaders: Rob Sternitzky and Kevin Clift | Indexer: J S Editorial, LLC | Publicist: Lindsay Cook | Marketer: Hannah Perrin

Manufactured in the United States of America

1st Printing

First Edition

The authorized representative in the EU for product safety and compliance is Penguin Random House Ireland, Morrison Chambers, 32 Nassau Street, Dublin D02 YH68, Ireland, https://eu-contact.penguin.ie.

To my animals,
you know who you are

Cities, like dreams, are made of desires and fears, even if the thread of their discourse is secret, their rules are absurd, their perspectives deceitful, and everything conceals something else.

—Italo Calvino, *Invisible Cities*

Contents

Prologue

Death and Desire

My heart is heavy with the things that I do not understand.

—Rudyard Kipling, *The Jungle Book*

You're in a building in Ghaziabad District, Uttar Pradesh, India. Time is moving at its usual languid pace, and there are three men speaking with the understated confidence of apex predators. The scene around them is prototypically human. Hallway. Flat ground. Thick concrete brick walls. Right angles. Windows placed at familiar distances, all of it lit up by warm afternoon sunlight spilling in from one side. There can be no threat here.

You watch the men walk confidently down this very quiet hallway, but you feel a murmur of discomfort. Then you realize what they're holding: tools, but the wrong kind for the place they're in. There's a long metal pole, which one man holds casually at its center, its two ends rocking forward and back with each step. Another of the three men holds a flat-headed metal spade, the edge set at a forty-five-degree angle—a tool humans have been using to till fields for almost four thousand years, made originally of animal shoulder blades. But it doesn't fit here: These men are wearing suits, and they're indoors.

They are talking, amped up, exuberant. You see nervousness passing over the face of one man. The faces of the others are turned away as they continue walking down the brightly lit hallway.

The first man reaches the corner and turns. He seems to be mustering courage, though he isn't quite succeeding. He raises the pole and shakes it, mimicking an action hero. The second man scurries a bit faster to catch up, swinging the metal spade back and forth as if trying to generate some momentum. He reaches the first man and the two of them face down the hallway, steadying themselves as they look at some menace beyond your view. You want to see it too. You approach, so close to the others that you can sense their jittery courage, this tactile energy that could spill in either direction: toward honor or cowardice.

The man with the pole taps it on the ground hard once, rallying himself and the two others. It makes a hollow clanging noise that resonates too brightly in the hallway. You're desperate to see what they see, scared of the terror but drawn to it. Suddenly, there's a rush, and you're being pulled back by the three men fleeing the unseen monster. They are not men anymore, though, but a crazed jumble of biological reactions. They grimace and yelp. Their legs pump and their feet scamper on the ground, trying to save themselves. They've been transformed into something new: prey.

You run with them, unable to muster the courage to face the creature head-on. Instead, you turn toward your left and stare at the wall as you flee, office windows and nondescript particleboard doors floating by you, each eliciting a pang of despair: None of these are exits. There's no way to hide fast enough when the threat is this eager to kill.

There is shouting behind you, the terrified sound of people being attacked. You risk a glimpse back. For a moment, the three suited men are standing in a triangle formation, two in the front and one between them a few steps back. Then for a split second it's as if they're enacting some elaborate choreography: They raise their legs, bend at the waist, and bring their hands to their faces in an overly theatrical gesture of protection. Then all at once they are thrown to the ground, the hollers now becoming screams, your vision scanning left and right so quickly that the scene becomes blurry: What is that shape moving through the tangle of men? Could that low supernatural rumbling, like thunder rending a hole in the sky, be the sounds of a living creature? The screams get louder. Everything is in the wrong place. Are those shadows? Is this beast readying itself to release you all into death? The chaos is rising, everything's a blur, and nothing makes sense anymore.

Quick fade to black. Silence.

The next thing you know, you're standing a couple feet outside of a jail cell. Inside the thick wrought-iron bars, a leopard sits on its hind legs, quiet now, calmly licking its chops, its eyes scanning slowly. It is no longer the malevolent force unleashed in the hallway. Now it is just a big, quiet, and beautiful miracle of creation, broken and compliant.

Suddenly there's a swipe of perfect royal blue across the screen, so very soothing after the fracas and terror. The blue is followed by the Platonic ideal of maroon, then purple, pink, emerald green, and jade. Finally, familiar words in a crisp white font emerge out of the blanket of colors. *Don't forget to LIKE and SUBSCRIBE*. The clip ends. It's called "Rampaging leopard with bloody mouth chases lawyers in courthouse attack." This is a useful title.

You reset to the beginning and this time the opening graphic, which you had glossed over the first time, hits different. WARNING: CONTAINS SCENES SOME VIEWERS MAY FIND DISTRESSING. You watch it again, this grainy film shot in the outskirts of New Delhi, and wait to feel something of that world.

▪ ▪ ▪

Living amid animals feels new, but it's an ancient phenomenon. Even as far back as 27,000 years ago there's evidence of a bargain struck between humans and a creature that has become among the world's most culturally influential. This was a period known as the Last Glacial Maximum, when ice sheets spread as far south as present-day New York and Hungary, permafrost blanketed Beijing, and glaciers sat atop Hawaiian volcanoes. It was also the era of the Pavlovians, an Upper Paleolithic hunter-gatherer society that lived in an area stretching from northern Austria into southern Poland, who navigated this cold and unforgiving time by hunting woolly mammoth, reindeer, and horse on the tundra. The Pavlovians made up a small and isolated community of humans, but they weren't alone. And the relics they've left behind provide one of the first glimpses into how human activity, even at a limited scale, can create long-lasting relationships with animals that influence not only how ecosystems function but also how culture emerges.

Since they were first excavated in 1924, the Pavlovian cave ruins have yielded an extraordinary cache of evidence of human predation on animals. This includes hundreds of thousands of large animal bones, hides, and hollowed-out fox teeth. But mixed within this boneyard is a curiosity that long puzzled archaeologists: well-preserved remains of ravens, *Corvus corax* (both words meaning "raven" and "crow"), the largest member of the Corvidae family and a bird known for its cunning and mystic powers. Deepening the mystery, it doesn't appear that the birds were a primary food source for the Pavlovians, because unlike other animal remains found at the archaeological sites, none of the raven bones discovered there have cut marks or burns, which would indicate that they were being used as a food source. So, if we weren't dragging them back to our caves for supper, what caused these Paleolithic ravens to flock to humans? After analyzing stable isotopes in the bones (isotopes are nonradioactive atoms useful for carbon dating and deciphering an animal's diet), archaeologists discovered that the Pavlovian ravens shifted their eating habits over time to more closely align with those of the cave's resident humans, feeding primarily on mammoth, reindeer, and other big game. Ravens have always been scavengers, following large predators to their next meal. During the Pavlovian period, an axial moment in human cultural evolution, the corvids seem to have switched their allegiance away from large members of the Carnivora family, like wolves, wolverines, and cave lions, and instead shadowed humans. It was an early bet on human ingenuity by a keenly intelligent bird, and an adaptation that reveals just how flexible the corvids are in exploiting the success of ascendant apex predators.

The Pavlovian raven bones are also evidence that humans had a profound influence on shaping ecosystems well before the dawn of history. Almost 30,000 years ago, before the rise of megacities, suburban planned communities, and inner-city collapse, and even a full 10,000 years before the Neolithic Revolution, a period of technological innovation that saw humans create the first permanent urban centers based around agriculture, our species was remaking nature in our own crude image. As the Pavlovians became expert big-game hunters, downing mammoths and reindeer herds with spears tipped with sharpened bones, they created a permanent "carrion landscape" of rotting carcasses around their caves and tent sites. This attracted a rogue's gallery of scavengers drawn to the mounds of putrid flesh, including the cave hyena and cave wolf, members of the Carnivora family that were big and scary enough to directly threaten the nascent human community. This forced the Pavlovians to curate the scavengers that fed on the flesh left within their spheres of influence, setting up warrior bands to force larger animals away while largely ignoring the smaller, faster, and more elusive scavengers like ravens and foxes, and instead allowing them unfettered access to our leftovers. Archaeologists have dubbed this landscape, with its permanent flesh heaps and small scavengers, an "ecology of fear," the creation of a whole new ecosystem shaped by the increasingly sedentary lifestyle of our species combined with our deeply rooted terror of sharing space with other predators. The upshot of this fear is that by forcing larger and more terrifying competitors to flee, the Pavlovians formed a stable, safe, and nutrient-rich environment for animals that threaten us less, setting us on a pathway to animal cohabitation.

There's one final step in this complex process by which humans reshaped our landscape and our animal interlocutors. With ravens and foxes always close at hand in Pavlovian society, their presence became, it appears, fundamental to the emerging culture of this early human social order. Ornamental raven feathers were used in Pavlovian rituals, while fox teeth—with small holes expertly drilled through—were assembled into pendants. These are early signs that as these animals gathered among us, they became not only physically close but also spiritually connected, and have been shaping our societies and beliefs ever since.

This is borne out by the remarkable spiritual power of birds across cultures. It's no coincidence that ravens and other corvids are among the most successful contemporary urban animals, with vibrant and growing populations scattered across the cities of North America, Eurasia, North Africa, and Australia. Most humans now live side by side with at least one species of crow or raven. While we force other successful urban animals, like rats, into the shadows, corvids are different. From a purely ecological perspective, it's evident that crows benefit from human behavior and habitats, but this reductive model cannot account for the cultural and spiritual ripple effects of being so close to such an intelligent, beguiling, and sophisticated creature. We may fear and even hate them sometimes, but they bring to us something rarefied that has been deeply embedded in our worldview over tens of thousands of years.

In the Norse pantheon, Odin had two raven companions who traveled the world and reported its happenings to him. Their names were Thought and Memory, fitting monikers for an animal that can distinguish individuals of its kind as well as their exact position in the dominance structure of flocks numbering in the hundreds. They also remember those of us who have wronged them: In a study of captive ravens in Australia, scientists gave the birds a piece of bread and then traded the bird's bread for a piece of cheese. In some cases, though, after receiving the bread back from the raven, a scientist would eat the cheese instead of trading it. Even after a month, ravens that were cheated refused to barter with the person who had cheated them and instead made exchanges with those who had dealt with them honestly. Among the Inuit, Raven was a god as well as a bird inhabited by a man, and it was Raven who created the world and all of the living things inside of it. Instead of leaving once his creation was finished, Raven could not shake his curiosity about animals and people, so he decided to stay and learn, just as his ancestors did among the Pavlovians. Ravens are symbols of filial devotion in Japan, perhaps reflecting their seeming loyalty to living with humans. It's in Japan too that ravens began to place walnuts in front of cars stopped at traffic lights, where the shells were subsequently crushed and their contents then eaten by the birds—a behavior that has been spreading across raven populations for the past fifty years. Among many First Nations, Raven is an expert thief, able to steal anything from anywhere, while elsewhere, the bird is a symbol of death and doom. Among the ancient Greeks, the father of Zeus was called Cronus, which has long been believed to refer to his role as "Father Time"; more likely, though, Cronus was based on the Greek word *corōne*, or "crow," on account of the belief that corvids housed the souls of ancient kings.

Ravens, evidently, have given our species something beyond an ecological benefit. They have imbued human organization with something much more profound: a spiritual bridge into another way of being in the world. It's a gift that urban animals are still giving to this very day.

▪ ▪ ▪

In 2007, Earth crossed an important threshold: For the first time ever, more humans were living in urban places than rural ones. It's perhaps no coincidence that the same year saw a social media platform,

Myspace, become the most visited site on the internet for the first time. These twin events reveal that while humans have become decoupled from nature, we still crave connection in any form. We yearn to be squeezed up against one another in space, and when that's not enough, we flock to be close online. Still, the city (or the internet, for that matter) isn't always a welcoming communal place, and it's easy to get lost in the glassy-eyed human gazes reflected back at us in grocery stores, in elevators, on subway platforms, in restaurants, at street corners, under the greasy fluorescents of dive bars, and beneath the ambient lighting of a pop-up art gallery / shoe store collab. It's the myopia of the city dweller: There are no animals here—at least none other than lapdogs, broken by training or genetics—and therefore no danger or spontaneity. All we see, anywhere, is us, staring back at ourselves with vacant looks. Nature, if it still exists, is something—somewhere— else. But that's not the real story. Look a little harder and a little longer, and there they are, all around us: the synanthropes.

You have likely never heard of synanthropes, though you surely know them. And they know you. Wherever you are as you read these words—in an office building, in a city park, riding the subway, lying on your sofa at home—you are surrounded by them. Synanthropes, derived from the Greek for "together with man," are wild creatures that have found a way to survive and thrive in human-modified environments. These are not our pets or our herds of domesticated animals but something like our oldest frenemies: beings from which we have long sought and failed to decouple ourselves, that now reside within our communities, and that have become an ineluctable part of the tapestry of our lives. They are the reason we hear birdsong in the morning and skittering in our walls at night, and why we take such pains to affix lids tightly to our garbage cans. But they are so much more than that too: soil churners, waste disposers, epidemic vectors, ecosystem partners, spiritual lodestars, and, sometimes, sharp-toothed marauders making their way through our most intimate spaces with cruel and bloodthirsty intent. These creatures are ambassadors from nature, arbiters of our planet's future, and a key influence on the continuing evolution of our species. Synanthropes have always been a part of our lives, in uneasy balance, despite our best efforts to feel insulated from the natural world. But that feeling isn't real. The fact is, we have never really been alone.

Take a walk in a forest and you'll be hard-pressed to spot an animal. Cities, though, are full of them, including some of the wildest and rarest on the planet, right there in front of you. That includes animals like the Superb parrot, its feathers an explosion of lime green unfolding behind it like fingers as it takes flight, a scarlet band encircling its neck, its head adorned with a creamy yellow cap, a creature so shockingly beautiful that looking at it feels as wrong as catching a naiad unawares; and yet, there they are, on the edges of parking lots, gorging on loose grain to the point of drunkenness, before wandering into traffic and being pressed into the dull gray concrete until their world fades to black. Or the grey-headed flying fox, a megabat as long-lived as a dog, with a body the size of a Chihuahua but far lighter, its binocular black eyes staring expressionlessly forward as it unfurls wings that dwarf its body—wings the color of tree trunks and as thin as paper, each notched with a single hooked claw, that stretch three feet across and carry it across the sky in search of fruit, dispersing seeds and pollinating as it tries to elude the city's death blasts: electrocution, extreme heat, barbed wire, shotgun shells, and endless other cruelties. There are others, many others, creatures on the brink of extinction yet wandering along sidewalks and footpaths that numberless humans tread upon every day. The tiger quoll, a three-foot-long marsupial carnivore that looks like a cross between a cat and a rat, with the grim temperament of a hungry grizzly bear and the world's second-most powerful bite. The collared delma, a tiny legless lizard that is neither snake nor worm though resembles both, uniformly brown save for a head that looks like a fat bumblebee. The loggerhead sea turtle, a marine reptile with a heritage stretching back 40 million years, which as a tiny hatchling must survive the claws of horned ghost crabs and the teeth of red foxes to make it to the safety of the sea. They are creatures that exist at the edge of fantasy, each so exquisite and rare that seeing one can change your life. And yet, there they are, these synanthropes, at decaying picnic tables, hanging from the trees planted along quiet city streets, in the bushes that line vacant lots behind squat suburban McDonald's franchises, and crawling determinedly across the sand between discarded plastic wrappers and melted ice cream cones.

Nature isn't retreating; it's accelerating toward us. A motley band of synanthropes, huge in number and diversity, are burrowing their way into our communities. In doing so, they are changing the places we call home. And they are changing themselves in the process.

We want to be near them because nature and its inhabitants lie at the very root of our spiritual selves. But we're also terrified, deep down in our bones, of their capacity to kill and bite and infect; we are scared of nature's cruelty consuming us. As our cities spread ever farther, and their effects ripple across the planet in disturbing new ways, we are backing into a strange new time. We have long lived with animal neighbors, having grown so accustomed to them that we go months or even years without perceiving them. When was the last time you really listened to sparrow song outside your window? What furtive shadows have you avoided at dusk? What sounds in your walls have you chosen to ignore? The sheer insignificance of synanthropes is among their greatest survival strategies. But that is just the beginning. The ways in which these animals have thrived among us despite our eons-long efforts to wipe them out is nothing short of miraculous. And as they've moved into our cities, synanthropes haven't just found new ways to live; they've found new ways to think. So how have they done it? And what creatures—larger, smarter, and wilder—will we be welcoming next?

This book is about how wild animals adapt to cities, and what we as city dwellers have done to force them into living with us side by side. The species that make up these pages—vertebrates and invertebrates, predators and prey—are wildly different, but they have all carved out unique roles in our urban ecosystems. That they have done so, in some cases with an overwhelming and terrifying level of success, compels us to recognize that our concrete-forward communities are legitimate natural spaces. But it is the remarkable adaptations of synanthropes—which span behavior, genetic mutations, the exploitation of human organization, and the creation of new social orders— that reveal how our cities have become not only hubs of biodiversity but also centers of rapid evolutionary change.

Part I will introduce you to synanthropes that live side by side with the world's greatest predator—humans—with each chapter exemplifying specific adaptations that they and others have used to fix themselves among us. Part II will show you what happens when synanthropes are too successful, and the chaos, danger, and fear that then arise. And Part III will reveal the ways human communities across the world are seeking harmony with synanthropes and imagining a future in which we transform our cities into places where species not only survive but are celebrated too.

Synanthropes show us the world as it is and a path forward to connecting with nature beyond simply mourning what we've lost. And if we look closely enough, they can show us that wonder is alive all around us, in the incredible adaptations of these creatures, our wild familiars, who have found solace in our streets, our garbage cans, our sewers, and our homes.

Part I

The Neighborhood Ark

Humans despair over our disconnection from nature, and cities have become emblematic of that deep existential crisis. Each new concrete pour and condo build feels like further confirmation of the inevitable end, somber epitaphs to the severing of our relationship with animals. The more we build our cities and alter the habitats that surround them, the harder it becomes to maintain that yearned-for interspecies intimacy. For this reason, ecologists have long referred to cities as "biological deserts," places devoid of the elements needed to sustain healthy and biodiverse ecosystems. Instead, it's believed, our kind dominates cities so completely that we have destroyed any possibility that other species might thrive there. This was taken as gospel for so long that most ecologists didn't even bother studying urban habitats, most presuming they would find nothing of interest there. By 2007, when for the first time in history more people lived in cities than in rural areas, only a few dozen scientific studies on urban ecology were being published annually out of a total of about 5,500 ecological studies, and it took until 2011 for the entire global field to muster more than 200 papers per year. It's as if, as far as most ecologists were concerned, cities were blank spaces on maps within which the laws of ecology ceased to exist.

But the staggering expansion of urban areas has made them impossible to ignore. And that has turned the notion of cities as biological deserts on its head. It's not just that ecologists have discovered cities harbor more wildlife than previously thought. It's that the animals living among us aren't just the usual suspects—pigeons and rats, insects and house sparrows—but, increasingly, are threatened and endangered species finding refuge in our midst. The startling truth is that cities are now, on balance, more biodiverse than the wild areas that surround them. As one recent study put it: Cities *are* nature.

How is this possible? It begins with the fact that humans are animals too, and like all animals, we are drawn to the most fertile, secure, and predictable habitats we can find. Cities, then, are most often built on natural habitats with high levels of biodiversity, where our species and many others are best positioned to survive over the long term. That's why, for instance, most urban centers are built on the shores of oceans, lakes, and rivers, providing us access to water, edible plants, rich soil, and an abundance of nearby prey. As our cities expand, choking out the surrounding areas, they subsume habitats that sustain species that can't survive elsewhere, and ultimately force those animals into becoming city dwellers. Urban areas have therefore become the last bastions of many of the world's species at greatest risk: 22 percent of endangered plants in the United States grow in the country's forty largest cities, while about a third of the world's cities harbor endangered birds. There's also a linear relationship, it turns out, between the likelihood that the home range of a species is urbanized and that they've been included on the International Union for Conservation of Nature's Red List, a systematic accounting of the world's endangered and at-risk animals and plants.

This all makes it sound as if synanthropes are passive creatures, engulfed by cities that doom them to extinction. But humans don't have a monopoly on adaptation. All around us, animals are finding that urban areas can sustain them, even if it means they must transform the way they move, speak, and even think. The city, it turns out, isn't just an ark to sustain creatures, but the best place on Earth to witness evolution in action. And it's all happening right outside your door.

Chapter 1

Puzzles

Our perfect companions never have fewer than four feet.

—Sidonie-Gabrielle Colette, attributed

Cities Expand the Animal Mind. Urban animals use a variety of tools to adapt to the dazzling complexity of cities, but cognition—the ways in which organisms process information and interpret the world—is among the most powerful. Like so many qualities of successful synanthropes, though, heightened intelligence isn't enough on its own. Instead, animals that survive and thrive in cities must sometimes transform the very way they think—a process that is unsettling how we understand our own evolutionary journey.

■ ■ ■

It was a brazen murder on the morning of July 9, 2015, in the heart of downtown Toronto. Was it a targeted killing? A hit-and-run? If there was a witness to the crime, they had long since fled. No one had dialed 911 for medical help. Fear—or was it apathy?—had made another life expendable, not even worth the brief phone call it would have taken to try to save him. And so, at around six in the morning, with a light wind stroking his face, the victim gazed toward the rising sun, obscured behind gray clouds, which were themselves blocked out by glass, brick, and steel monoliths casting shadows that inched slowly toward him as the light spread through the sky. And it was then that he yielded to the ineffable, unrelenting momentum of death. He was just another unremarkable victim in the heart of the city. Afterward, all that remained was the body—huge, by all accounts—splayed on the corner of Yonge and Church, a busy commercial intersection. Here, in this diverse city of almost 3 million people, long considered a haven of safety, equality, and community, with a murder rate among the lowest of big cities in the world, the body was an aberration.

It lay there throughout the morning rush in a shocking and obscene pose, naked, supine, legs spread apart shamelessly, thin purplish-brown lips pulled into a rictus. The mouth was agape, the tongue like a hard, blunt arrowhead held between dirty teeth, the hands hanging limply from stiff bent wrists. This body, a reminder of the basest laws of nature, in contravention to the lies we've swallowed that life is static and society's rules will protect us at all costs—it was this body that transformed the city.

Noon. The sun had climbed in the sky, its diaphanous rays spilling gently onto the metal, plastic, and concrete admixtures that make up downtown Toronto. Filaments of light, finding narrow paths between the patchwork of summer clouds, covered the city like a cobweb. Six hours after the death, workers began filing out of local office buildings for lunch and noticing the body, pressed against the concrete curb, impossible to miss in the rising summer heat, its flesh now stiff and emitting a gentle odor of decay. Some passersby stopped in shock but were soon pulled back into the pedestrian flow sidestepping the corpse, everyone too preoccupied with their own lives to take on yet another problem.

From a second-floor tech start-up across the street, four men peered down with a mixture of bemusement and horror. How long would the body be left to rot? Who—or what—was responsible for taking it away? And how uncaring did a city have to get before its spirit died? The men, all in their twenties and thirties, didn't have the skills to solve the problem of an enduring corpse. But moved by the moment, they mustered something that none of them realized they possessed: an innate sense of decorum.

One of the men visited a nearby flower shop and purchased a single rose wrapped in cheap cellophane. Another headed to a nearby pharmacy to get a card. "The warmest wish / That I could send," read the printed text on the card, "would only begin to show you / how much / I wish you good health." Underneath, the men had added their own message: "Hang in there! ♡ The Gang." They brought the gifts to the corpse, sliding the card under its left foot and placing the rose in the crook of its left armpit, careful not to touch the body itself in deference to the deep biological imperative to separate the living from the dead.

Satisfied, the men returned to their office to renew their dedication to code writing, data security issues, email communication, and internet consumption. But having pierced the ontological plane, they found that reality had been transformed. Before the card and the rose, the victim was an anonymous corpse. Now? A cult figure in the making. After a few minutes, the men peered back out the window and, to their surprise, saw that the body was now obscured by a crowd.

People were taking photos, phones bent at impossible angles, the victim given a second life online in death. The card filled up with messages. "RIP beautiful soul! ♡♡♡"; "Hope it was quick! I miss ya!"; "♡ Ruby"; "Rest in peace"; "Sorry buddy —Sara"; and finally, just three simple letters that seemed to sum up the complicated feelings of all those who were coming to pay their respects: "BYE!"

By 2:15 P.M., news of the corpse and the makeshift memorial had spread across the world. A framed photo that looked uncannily like the corpse itself was added to the memorial. A votive candle appeared, its flickering flame imbuing a sense of the sacred. A cigarette appeared in its hand. The crowd had even honored the victim—still unknown, just another John Doe in a city full of them—with a name: Conrad.

Online, images of Conrad and his spontaneous wake were like some new disease: highly transmissible and slightly gross. And that is how, on a defining day for the city of Toronto, a simple motto was born: #DeadRaccoonTO.

Day collapsed into night. One candle became many. The crowd became so dense that people began to organize themselves into a line to write condolences and pose for photos with Conrad. A waning crescent moon took its place in the sky and a handful of stars shone down weakly through the light pollution that covered the city like a weighted blanket. And still the people came to pay their respects. It was as if the dead raccoon had become a midsummer version of a mall Santa: fat, clever, covered in fur, and obsessed with sneaking into people's homes.

Finally, mercifully, at 11:23 P.M., roughly seventeen hours after his death, a white Toronto Animal Services van languidly drove up, stopping four feet from where Conrad lay. With the votive candles lending the scene a quiet solemnity, an unsmiling sanitation worker descended from the passenger side holding two extra-large black garbage bags. A small crowd remained, watching as the worker brusquely grabbed the corpse and, careful to avoid touching it directly, wrapped it in the garbage bags, then threw it into the van. The crowd gasped in horror; the sanitation worker looked up, clearly exasperated by the scene. "Seriously, man?" he asked with barely concealed disdain. "It's a dead raccoon."

Within ten minutes, the deed was done. It was now 11:35 P.M., and all that remained of the drama was the cigarette, the now-wilted cellophane-wrapped rose, an empty donation box, and an extinguished votive candle, all of it just trash on the ground. "Damn," read the last #DeadRaccoonTO tweet of the night. "Life's so short."

■ ■ ■

If you ask anybody in Toronto, they'll tell you that raccoons, a.k.a. the *Procyon lotor* (Latin for "before-dog washer," given their apparent penchant for washing their food), are everywhere. The creatures have turned even the most gray urban spaces into wild landscapes, which have come to suit them far better than their original woodland home. There is no part of the city that they can't master—or at least that's what people here believe.

I want to test that theory by going to the most urban, concrete, and dead place I can think of to see if I can catch a glimpse. To my mind that's Union Station, the epicenter of the city's rail and transit system and the doorway into its sprawling and grid-like downtown district. If I see raccoon traces there, I figure, it will be pretty good evidence that these creatures have unlocked even the least accommodating micro-habitat across this massive metropolis. So I take the subway during rush hour, just as a raccoon did a few days earlier (after which it was praised for being so well-behaved), and emerge amid a crush of human bodies.

Union Station is surrounded by concrete sidewalks, made glossy by old flattened gum, which are in turn ringed by downtown streets filled with snarled and unmoving traffic spewing clouds of choking exhaust. All around, gleaming with an incandescence that seems cut out of the cold blue sky itself, sheer cliffs rise impossibly high, a canopy of skyscrapers reflecting sunlight among themselves, creating a labyrinth that partitions the sun's rays.

At the epicenter of this hyper-urban space, I walk through Union Station's broad galleries, taking stock of the throngs of human experience passing through. At one point, the light shifts to a warm buttery glow, and I look up at frosted skylights etched with a branching mosaic of semicircular blobs surrounded by droplets that spread out and overlap across the panes in a crude motif. There's something about the pattern that causes me to pull away from the stream of humans and set my back against the wall. The blobs and droplets appear randomly placed at first glance but seem to have an internal order the longer you stare. It's a surprisingly organic, aesthetic choice for a space that is otherwise made up of straight lines and antiseptic materials. I like it. I follow the rough etched lines across the opaque glass to the corner of one of the skylight panes, where a single semicircular blob and set of droplets—five in all, I see now—are etched apart from the others. And it's then that I realize what it is I'm looking at.

This isn't a pattern etched on glass. Instead, it's a layer of dirt and grime built up over seasons, with prints made by the press of raccoon paws. The semicircular blobs are from the carnivoran's metacarpal pads; five digits extend out, short like a dog's paws but spread wide like human fingers. I'm floored. While a nonstop parade of human beings moves through this space every day, raccoons numbering in the hundreds are moving back and forth among this static urban architecture above us. Here, in a patch of the city that is 99 percent concrete, a space that leaves no quarter for animals to survive, is evidence that the creatures are thriving, albeit just out of reach.

I circle the station many times, then move surreptitiously through the buildings across the street to catch glimpses of the community living on the roof. But no such luck. Later, I head over to a friend's third-floor apartment, only to hear that I just missed an epic battle between their cat (Kenneth) and the slow-moving resident raccoon that lives at the top of the fire escape. It's as if raccoons only exist in my peripheral vision. Like distant stars, once you try to look at them straight on, they disappear.

In the weeks that follow, these frustrating near misses pile up all across the city. Eventually, I realize I need professional help.

▪ ▪ ▪

When I meet up with Rob Gordon at Cafe Paradise, a brightly lit coffee shop a few blocks from his place, he's looking slightly worse for wear. It was a long night: between midnight and two in the morning, seated by the back door of his neighbor's apartment next to his landlord's cat (Jason), he had weathered an extended siege by a mother raccoon and her four kits, who were intent on breaking through the cat door and into the kitchen for scraps. Jason was hissing fiercely;

Gordon, who projects a monastic calm, was trying to reason with the mother while batting her back each time the flap swung open. "I spent one hundred and fifty dollars on Amazon last night," he complains, "buying coyote urine as a deterrent."

Gordon is cherub-faced, with low affect, speaks in a laconic and analytic style, and has a wry and elusive smile. The times I hung out with him he wore black and dark blues exclusively, his long brown hair tied up in a ponytail that ran down to his lower back. Gordon is an exterminator—actually, a pest control expert—who's been in the game for a decade. He's also low-key one of Toronto's most important musical operators, having toured for years as a drummer with the world-renowned violinist and composer Owen Pallett (who first rose to fame as a member of Arcade Fire), while also having started some of the city's coolest venues, which at times doubled as his home. "At one point," Gordon says, "I was living on airplanes, never seeing the sun, and working until five A.M." He ran through about ten different jobs, including barback, line cook, construction worker, server, and random shift worker, which he could jump in and out of depending on his tour schedule. One day his ex-girlfriend's roommate, the manager of an arts space called Studio Gallery, asked him whether he wanted a quick and well-paying gig. (Gordon knew him from the arts scene; he had thrown an epic Daft Punk after-party at the gallery a few years back to which the duo even showed up.) A year earlier, one of the gallery's artists in residence had dragged a bedbug-infested couch off the street and into the space, and it promptly infested the building; instead of hiring someone, the manager decided to deal with it himself and, buoyed by his success, started his own pest control company. It didn't take long for Gordon to realize he also had a knack for the work, and he quickly became a top lieutenant in the nascent company. Still, he wasn't quite ready to fully commit to pest control as a vocation. That is, until an incorrectly administered injection left him with permanent paralysis in one arm—a potentially career-ending injury for a drummer. After that, Gordon was forced to withdraw from performing. As he thought about the things he loved in life—his wife and daughter, mostly—he realized it was time for him to build something that could sustain them all. So he decided to strike out on his own. "I'm allergic to money," he says. "So do I want to be running a pest company? Not really. I'm not even interested in pests," he adds, "but I am interested in people." Gordon calls his company Out of Body Pest Control.

When I ask Gordon about raccoons, he sighs. "I'll see ten dead raccoons on the way to a job," he says, "practically every day." It obviously hurts. Gordon respects the animals' capacity to solve whatever puzzle they encounter, but it's their vibe that he appreciates more than anything else. "When I trap a squirrel," he says, "the energy is insane. They'll bang their head against the bars of a cage for hours until the fur is completely stripped and their head is bloody. I ride around with them screaming at me from the back of the van." Raccoons are different. "When I catch a raccoon," he says, "ninety-five percent of the time, it looks like a teenager that's been busted for smoking pot: just kind of aware of what they've done and accepting the consequences." He laughs softly. "It's not like they high-five you on the way out, but they're so bashful and calm."

▪ ▪ ▪

In one of the most diverse cities in the world, which welcomes tens of thousands of newcomers each year, a run-in with a raccoon (or raccoons, for they rarely travel alone) is a sign that you've truly arrived. The raccoon is a member of the Procyonidae family, named after Sirius, the "Dog Star," which is the brightest star in the constellation of Canis Minor. Dome-shaped, like a furry armadillo, and ranging from the size of a large cat to a small border collie, raccoons sport a mask of black fur around their eyes surrounded by a halo of white, set above long thick whiskers. With rounded bear-like ears, a thick black-and-white ringed tail, and long front arms that end in finely tuned paws, they look like the Hamburglar in a fur coat. They're also just as hungry and as cunning, and can weigh up to thirty-five pounds.

Raccoons are indeed everywhere in Toronto, a by-product of urban sprawl, which has seen their ancestral ranges become absorbed in a blink of ecological time into human habitats. Rather than flee or perish, though, the *Procyon lotor* has turned out to be uniquely adaptive to city life. Estimates have put Toronto's raccoon population at an absolute minimum of 100,000 and as many as 640,000—an astonishing ratio of up to one raccoon for every four people who live here. And this explains why practically every Torontonian has a raccoon story. These include monthly power outages sparked by raccoons fatally attracted to the warmth of electrical power stations, which regularly affect tens of thousands of households. Or the time a person flipped their car over to avoid hitting a raccoon that was casually strolling across a busy downtown intersection (neither the driver nor the raccoon were injured). Or the time a raccoon, described by staff as a "perfect client," followed a customer into a busy McDonald's, sidled up to the counter, and refused to leave until it was given a Chicken McNugget. If New York City has 8 million stories, Toronto has just as many—but they're all about some crazy raccoon.

It's not just that there are a lot of raccoons in Toronto. These members of the order of Carnivora, which includes bears, wolves, felines, seals, and many other mammals that specialize in eating flesh, have found a way to reorganize themselves to take full advantage of all that city life has to offer. In forests, their population density reaches a maximum of about thirty-six per square kilometer. In Toronto, some estimates suggest there could be around one hundred raccoons per square kilometer, meaning their population is about ten times more dense than in their home ranges. It's an ungodly crush, so clearly out of balance that it's as if the city has been designed as a veritable paradise for the mesopredator. So, how could a raccoon, an animal that evolved over hundreds of millions of years to adapt itself perfectly to swampy forests, find such runaway success living amid the most destructive and systematically ruthless species ever to exist?

Their success lies in co-opting human design. Raccoons, like humans, approach new environments with a mixture of caution and curiosity, reflecting their innate capacity to understand multiple threats and opportunities simultaneously. This effortless multidimensional thinking is linked to two major traits—cognitive flexibility and altered home ranges—that give synanthropic raccoons a unique leg up among the city's menagerie, and provide tantalizing hints that cities might be changing the way the animals think.

Cognitive flexibility refers to how "plastic" an animal's thinking can be, or how well it can find new ways to solve problems when confronted with unfamiliar situations or environments. Scientists test animal cognitive flexibility through what's called a "reversal learning" test. In one recent study by the U.S. government's National Wildlife Research Center, scientists trained raccoons, skunks, and coyotes in Colorado and Utah to expect a reward (pellets of chicken liver or sausage) if they tapped on one of two glowing buttons stationed at the left or right of a mocked-up food dispenser. The first stage was simple: If the button on the left was pushed, food pellets would drop from a hole. Compared to skunks and coyotes, raccoons were the quickest to discover that tapping the left button would release food, with most of the eight raccoon subjects unlocking the puzzle within one or two attempts. Skunks lagged slightly behind, needing four or so attempts before they got the hang of the new regimen. But coyotes—anxious, clumsy, and scared to try new things—lagged far behind their mesopredator cousins. It took forty-four tries before a lone coyote, named Orion, pressed its paw down on the left button. The other five coyotes simply refused.

This was just the beginning. After weeks of having it drilled into their small brains that tapping the left button equals food, the scientists abruptly shifted the environment. Suddenly not only did tapping on the left button fail to provide food but also it initiated a time-out phase in which the dispenser stopped responding for ten seconds. Getting food now required that the animals unlearn what they had just been taught and go against their newly primed instincts by choosing to tap the other button, on the right. This was when the cognitive flexibility and raw intelligence of raccoons truly shined.

Raccoons were aces at recognizing when their newly acquired skills became useless, and they quickly pivoted—far faster than the skunks or coyotes—to experimenting with new strategies, including hitting the button on the right. This phenomenon is called a "paradigm reversal," an apt term for the kind of confusion and recovery that raccoons (and sometimes humans) experience in urban spaces every day. And while it's rare that a raccoon might encounter a food-dispensing button in its nightly travels (probably as rare as a human stumbling across an ATM dispensing free bills), the capacity to take advantage of these kinds of opportunities, even when they mean going against the ostensible safe bet, is a key reason for their near total dominance of cities like Toronto.

The swampy forests that raccoons and their ancestors evolved within for 28 million years couldn't be more unlike city landscapes. From the perspective of a single raccoon, any patch of forest is a familiar mixture of trees, plants, fungi, ponds, and animals organized into different combinations, with so many variations that it would be impossible to map. One level of magnitude higher, though, and the forest is a monolith. In a healthy forest, each section pretty much resembles any other. Where differences between forest sections emerge—rocky outcrops, ponds, or grassy clearings—they do so gradually. Mud and long grasses signal a water source nearby, while a steepening hill beset with pebbles and stones warns of an adjoining cliff face. Features rarely emerge out of nowhere.

Cities upend this intuitive organization. Any individual, be they raccoon or human, could quickly learn to navigate a single section of an urban environment. Parking lot. Street corner. House basement. Apartment building roof. But one level of magnitude higher, and cities—unlike the monochromatic green of a verdant forest canopy—disassemble into a chaotic and seemingly irrational patchwork of microenvironments. There is no inherent relationship between the vertical cliff face of a Walmart's exterior walls and the electronics counter, McCafé, displays of fruits and vegetables, and passport photo booth located inside, all illuminated under the blinding wash of fluorescent bulbs held aloft three stories up. That sheer cliff, in turn, has nothing to do with the area that surrounds it: a clearing of poured concrete and painted yellow lines populated by constantly rolling metal vehicles moving at different speeds in multiple directions alongside human beings making their disordered way in and out of their cars and the building. Extend it farther, and there is no intuitive reason that the parking lot abuts another flattened concrete-covered plain, this one with vehicles moving in two directions at much higher speeds. Beyond the road lie much smaller buildings organized in neat rows made up of a jumble of hard angles—porch steps, open windows, sloping roofs—that seem to spring spontaneously from the ground. Each one of these fragmented microenvironments—Walmart, parking lot, road, residential block—are found side by side. From an ecological perspective, it doesn't make any sense at all.

There is very little about finding food and shelter inside a Walmart that is translatable to the parking lot outside. Moving from one to the other is the equivalent of finding that the button to the left shuts down the system entirely, leaving you starved. Mastering the laws of the parking lot can't possibly prepare you for the throttling speed and amoral violence of the road beyond. With cars driving down roads like unpredictable boulders, disobeying the most basic laws of spatial reasoning, the time you have to unlearn the rules of the game is limited if you are to survive. And so it is for the houses beyond, the park beyond them, the skyscrapers, corner stores, restaurants, bars, alleyways, hospitals, schools, and construction sites that lie beyond, beyond, beyond. Mastery of one of these spaces briefly staves off death. But mastering the constant process of learning and unlearning as you move from each to the others—now, that is a radical adaptation, and it's a rare synanthrope that can manage this feat. Cities have revealed that raccoons have the power to reset their internal logic and learn the game all over again each time they enter a new urban microhabitat, transforming this unremarkable swamp dweller into a model urban thinker. And the strangest part of all? We have no idea how far their minds might go.

▪ ▪ ▪

It's almost 8:30 P.M. in Tompkins Square Park in New York City on a hot July evening. My attention is wholly fixed on a pair of off-brand wireless earbuds, which will serve as my gateway into an ecosystem hidden in plain sight. I pull my phone out of my pocket, a pinprick of bright blue fluorescence shimmering across the hexagonal concrete tiles surrounding the grand elm tree at the center of the park. Here and there the light is absorbed by crude rectangles of worn-down black tar, the city's flattened scar tissue, which absorb luminosity like celestial anomalies. And then it's time. The susurrations of the humans surrounding me disappear as I put my earbuds in, the real world blocked out by the prerecorded sounds of this exact same place at another point in time. It's a kind of auditory disguise—or rather, self-guise—designed to make me feel as if I am where I actually am: Tompkins Square Park. A few seconds later, a soothing female voice broadcasts directly into my ear canals.

> Why do we hate them so much? . . . Look around.
> Do you see any rats?
> Try to look for rapid movement near the ground.
> They might be hard to spot, but their nest is nearby.

This is "The Synanthrope Preserve: The First to Cross," a thirty-five-minute audio tour led by the syrupy-smooth voice of a narrator intent on making me make contact and ultimately fall in love with the park's resident rats. I don't have to wait long. As I scan the ground in front of me, I see the first rat running along the curve of the concrete footpath, silently moving through alternating patches of weak shadow and manufactured light, its rough coat shifting from dark brown to silvery gray. Then comes the second, hopping between two fenced-in clumps of foliage lining one of the winding concrete trails. The third and fourth skitter past each other in the underbrush near the dog run, either fleeing some unknown danger or each other. Obeying the narrator, I turn up the winding path again toward the grand elm at the center of the park, and that's where I see the fifth rat, making its way slowly, ponderously, its left shoulder grazing the raised brick at the path's edge. The sixth I have no recollection of, except that it flashed a creamy belly. I see the seventh, my final rat, as I exit Tompkins Square Park at the close of the audio tour. It scurries across the path in front of me, loping and bounding from the direction of the street toward the relative safety of some stunted, sickly looking bushes next to a chain-link fence. In its mouth it holds a perfect, impossible prize: a golf ball–sized sphere of orange food so bright that it seems to glow.

Rats even purr when they're happy.

I am happy for the rat—proud, even. With each of the seven rats, I feel a quiet thrill, my heart leaping as the promise of connection is fulfilled. The narrator in my ear has succeeded in radicalizing me. I want the rats to win. I want the rats close. I want the rats to crawl on me and let me caress them.

Along with humans, rats are the world's most widespread urban mammal. In New York City, rats have become such a fixture that it's no longer clear how the city's ecosystem would function without them. And while we know a lot about rats as carriers of plague and disease, and that they systematically weaken infrastructure on account of their endless, cheerful gnawing, the features that make them so essentially ratty have largely been ignored. But it's impossible to ignore rats forever: As much as they would like to be left alone, their alarming success means that humans can't help but stumble upon them.

Toronto's raccoons have made stunning cognitive leaps to master the city, their minds evolving, it seems, to meet every new puzzle they encounter. In New York City, the brown rat, *Rattus norvegicus* (Norway rat), has long been caricatured as a kind of super-pest, but it's not until recently that we have begun to understand how the pressures and opportunities of the city have led to profound changes that go far beyond cognition. New York City rats aren't just well adapted; they are the product of genetic changes brought about during their time living, eating, mating, and mutating amid human activity.

The brown rat branched off from its most common recent ancestor between 1 and 3 million years ago, but its evolutionary success was never assured. As recently as fifty thousand years ago, the rat was experiencing a major population decline and seemed at risk of extinction. This trend was reversed only about ten thousand years ago during the Neolithic Revolution, when humans shifted from living in small hunter-gatherer groups to larger agricultural communities, where sudden surpluses of food gave the rodents a lifeline.

Hardly any brown rats exist outside urban areas, but they seem to have only spread from their ancestral range in northeast China in the last millennium as the human population expanded across Asia and Europe. Using the remains of rats culled from shipwrecks, archaeologists have dated the introduction of the first brown rat into North America to the year 1731. That makes New York City the scene of roughly three hundred years of rat evolution, which has resulted in a breed of rat the city can truly call its own. Compared to brown rats still living in northeast China, those in New York City think differently, move deliberately, and are just plain badder. And when scientists mapped the genomes of brown rats in New York and compared them to their northeast Chinese brethren, they discovered that many of the changes that turned NYC rats into such formidably successful synanthropes likely happened within the past half century or so, suggesting that the city is causing rats to evolve at an unprecedented pace.

NYC rat adaptations come in four basic flavors: the way rats metabolize toxins, what they're able to eat, how their minds work, and how they move. Starting in the 1950s, the New York City Department of Health launched an aggressive rodenticide campaign using warfarin, an anticoagulant that causes rats to bleed to death. But the local rats evolved to tolerate, and then fully ignore, warfarin, forcing the city, in the late 1970s, to roll out a second-generation anticoagulant rodenticide—which became just as ineffective as the first one within a matter of years. It's a bit like humans somehow becoming immune to drinking gasoline within a few generations. But that's just the start. New York City, like many other North American cities, is rife with polychlorinated biphenyls, or PCBs, a highly carcinogenic class of chemicals originally used, among other applications, to keep fridges cool. Though they were banned in 1978, PCBs are forever chemicals, and they're frustratingly difficult to remove once they've contaminated soil, sediment, or waterways. Once absorbed by a living creature, they bind tightly to fatty tissue and cause cancer, liver damage, hormone disruption, skin sores, and even neurotoxicity. But what does the New York City rat care about PCBs? Not a thing. In the decades since the 1970s, the city's rats have undergone a genetic mutation conferring PCB resistance, meaning they're able to remain as healthy as an urban rat can.

Beyond these superpowered mutations, rats in New York have undergone subtler changes that allow them to weather the unique stresses of city life. Chief among these is a mutation in *CACNA1C*, a gene associated with a cluster of psychiatric conditions, including autism, schizophrenia, depression, and chronic anxiety. The expression of *CACNA1C* is related in part to childhood stress, which leads to the obvious question: What could be more stressful than being a young rat in New York City? From its earliest waking moments, a NYC rat pup must deal with flickering multi-prismatic light at all times of the day and night, loud booming noises out of nowhere, massive predators at every turn, furiously speeding metal kill machines, and an environment so counterintuitive and spatially disconnected that it crosses into the absurd. Those stresses, introduced early in their lives, are epigenetic factors that appear to switch *CACNA1C* on. Consequently, rats with the gene expressed turn into loners, less interested in social play like wrestling and pinning. When they do socialize, these loner rats tend to be rougher, pinning down other rats for longer to get the same level of dopamine that others do with much less social play. Like the city's class of burned-out partygoers, NYC rats must go to extremes just to taste a rush.

They look different too. If you're lucky enough to run your fingers along the skull of a New York City rat that lived in 1890 and compare it to a modern-day rat from the city, you'd be amazed at what 135 years of urban living can do. As New York moved from eighteenth-century modernism to the twenty-first century's culture of despondence and replication, the noses of the city's rats grew noticeably longer, an adaptation believed to be better at warding off the cold of a hard snowy city. But it's not all hardship. Over a century, as the food industry transformed snacks into science experiments, rats got in on the fun, growing fat on the foods we grow fat on, like pizza, ice cream, mayonnaise-soaked soft bread, and Xxtra Flamin' Hot Cheetos. In the case of the rat, this diet also seems to have reversed 50 million years of evolution. Back then, all rodents—the great ancestors of *Rattus norvegicus* included—had short stubby molars designed for grinding soft fibrous tissue like plants and scavenged flesh. As the buck-toothed rodents expanded their diet to nuts, bone, vegetables, and seeds, they also, over eons, developed taller molars less prone to damage from these tougher foods. But New York City has, it seems, slammed the brakes on this long-standing evolutionary trend. Rather than growing longer teeth, the city rats' molars are instead evolving in reverse to again become smaller. The reason? The decadent foodstuffs of modern urban life are as soft as their diet from 50 million years ago. Rats like pizza as much as we do, it seems—if not more, given that they've evolved to consume it as efficiently as possible.

Meanwhile, in an era when real estate prices are making living in the city a dream for most, data from the Department of Health's Rat Information Portal shows that rats are mostly congregating near subway lines, schools, and restaurants. New Yorkers may be doomscrolling Zillow, fantasizing about historic brownstones in upscale neighborhoods with high walk scores. Rats don't wait for their dreams to come true: They just move in.

And when they do, they don't leave. Once ensconced in a neighborhood, one recent study found, the city's rats remain fiercely loyal to it, so much so that they have broken into genetic splinter groups that differ from others only 650 feet away, which is the maximum boundary that one generation of rats will generally travel within (though most only travel about 100 feet). These inter-cluster differences are so significant that, over time, they have led to two genetically distinct evolutionary lineages: uptown and downtown rats.

The audio tour of Tompkins Square Park heightened my senses to the furtive movement of rats. I felt alive to their industry, their dogged determination to live, and the humility they possess in building a world so close to our own. In just a few short minutes, my feral aversion to rats was replaced with something bordering on love. Evidently I'm not alone. On the inaugural night of the audio tour, a crowd of hundreds showed up to come face-to-face with one of the city's most reviled creatures. In a town overflowing with rats, it seems, humans are still desperate for communion.

■ ■ ■

A week after Rob Gordon and I first met up, he sends me a text.

> Hey Dan. I'm picking up a raccoon from behind paradise cafe right now.

It's the same spot in Toronto where we met the first time, and I quickly bike over to meet him in the alley. When I arrive, Gordon is sitting at the wheel of a parked van, casually flipping through his phone. "Wanna see him?" he asks. I nod, barely concealing my excitement. He walks to the back of the van with a weary demeanor and pulls up the tailgate door.

Quiet, calm, and caged, a male raccoon, his fur glossy and glistening in the midday sunlight, slowly looks around and considers his circumstances. He's a prisoner but betrays no anxiety, just a mild plaintive look accentuated by the peaks of black above his eyes, which extend diagonally down his cheeks like two long splashes of expensive Japanese ink, and which give his diamond-shaped face a searching expression. He's a handsome fellow, lean and healthy-looking, lying on his belly with his forepaws extended, possessing the regality usually reserved for stone lions. His pate is silver, his whiskers white and lustrous, his body the color of straw rather than the inky black and silver shading I'm used to seeing on the carnivorans. I can't take my eyes off him; in return, he looks through me—that gaze—and then beyond. There is no exchange of fear between us. He knows what I am, what Gordon is, and is patiently waiting for whatever comes next, a creature utterly at peace with fate.

Gordon pulls the door back down. "Ready?" he asks. *Not quite,* I think, but no matter. Soon, we're driving through grinding traffic on our way to High Park, a four-hundred-acre semi-wild green space in Toronto's west end. "You can kill pigeons and you can kill squirrels,"

Gordon explains en route, "but you're not allowed to kill raccoons." That has turned High Park into an unofficial dumping ground for any raccoon unlucky enough to get caught. As we turn down Parkside Drive, Gordon flicks his head. "Dead raccoon," he says as we pass a disassembled mass of dark fur spread across the curb. A few minutes later, we turn into a quiet parking lot surrounded by trees in the south end of the park, close to Lake Ontario. Gordon, for the first time since we've met, is on edge, scanning the parking lot for passersby. "You never know how people are going to react," he says, a bit guardedly. A family walks slowly past, oblivious, and we wait it out. Finally satisfied, Gordon opens the back door of the van and pulls the caged raccoon out and onto the dirt at the foot of an outcropping of trees. "Check this out," he says. "As soon as I open this door, he'll rush out about five feet, climb up a tree, and then turn back to look at us." He opens the cage and, as predicted, the raccoon immediately dashes away and up the back side of a nearby tree, then cocks its head from behind the trunk, looking at us with a singular feral intensity. Gordon looks at the creature with a mixture of benevolence and mild interest. A few seconds later, the raccoon hops down and disappears farther into the forest. "See?" Gordon says as he slams the rear door shut and we climb back into the van. "Sometimes with the real fat boys, they need to waddle out the cages backward because they can't turn around. They get pretty well-fed."

We start back the way we came, and I ask him whether the work is emotional for him. He ponders for a moment. "It's a really particular experience going up into a dark attic and shining a light, and you know there's probably something in there," he says. "I've had to remove dead raccoons, and it's pretty gross. They start to leave their form; it all starts to fall apart." As we drive, Gordon leans into his war stories. Once, a Hasidic woman laid out her collection of wigs and asked him to spray them all down with toxic chemicals to remove a flea infestation. She was insistent, but he declined. Not only would it have been dangerous; it's illegal too. Another time, in one house in Rosedale, Toronto's toniest neighborhood (made famous on the cover of Kendrick Lamar's epic diss track "Not Like Us"), Gordon entered a mansion filled with thousands upon thousands of flies but no smell; he eventually pried up some floorboards, where he found the desiccated corpse of a raccoon that had found its way in but couldn't get out. "Anybody can have pests," he says. "They don't discriminate. It's very equalizing."

▪   ▪   ▪

Strong emotional bonds are hardwired in raccoons. In the wild, though, the urge to connect mostly stops at the family unit. Woodland raccoons have home ranges ten times larger than those in urban areas, with far lower population densities. That's because natural habitats generally contain smaller, sparser, and variable food sources, while those in cities are large, stable, and clumped closely together. One study found that, across the province of Ontario, some rural raccoons regularly traveled as far as thirty miles, while those in Toronto, the province's capital, rarely traveled more than three city blocks. With as many as one hundred raccoons per square kilometer in the city, one of the urban features raccoons have had to adapt to is other raccoons.

In the place of natural intraspecies aggression, raccoons have developed new ways of being with one another that have profoundly shaped their success in cities. For a long time, biologists believed that all raccoons—rural and urban—were only ever loosely connected, with males more likely to engage with males outside of their family unit, and then only seasonally. But newer research has shown that urban raccoon groups, which are known as gazes, have far more complex and durable social systems than previously thought. At minimum, it's now believed, a single raccoon will interact with 50 percent of all other raccoons in its home range. In Toronto, where the animals have home ranges stretching about three city blocks, each raccoon is on average regularly interacting with about ten others outside of its own gaze, while contact between gazes is likely higher in areas of the city where the animals are more tightly clustered. Across the entire city, that puts tens of thousands of raccoons at only one degree of separation, all of them teaching and learning from one another. Those tight networks arise, biologists say, because urban areas are full of large open spaces—apartment building courtyards, leafy parks, alleyways linking multiple backyards, and the roof of Union Station—that accommodate massive dens in which multiple families, and as many as hundreds of individuals, live together.

Gossip is a dagger that destroys; it's also the glue that holds gazes together. For raccoons, urban communal living brings the added benefit of hyperefficient communication networks. When one wily raccoon learns how to unlock a new food source, it's impressive. But in the forest, the ingenuity of a single animal is more likely than not destined to be forgotten when it dies: There just aren't enough other raccoons that have been exposed to its tricks. When one urban raccoon discovers how to undo the locking mechanism in a new generation of raccoon-proof compost bins, that competitive advantage is rapidly diffused across their family, their gaze, their extended social network, and beyond. And that means that when a lone raccoon experiments with a new tactic and unlocks a universe of food, it's only a matter of time before others will follow suit. A singular moment of genius becomes a step increase in urban adaptation.

In Toronto, this isn't just hypothetical. In 2016, after years of raccoons having their way with compost bins, the local government unveiled a thirty-one-million-dollar solution: a next-generation green bin with a rotating locking mechanism on top billed as raccoon-proof. "Raccoons apparently don't have thumbs so they don't have the ability to turn a lever," Toronto's director of collections and litter operations explained at the time. "There's not a raccoon that's gotten into it yet," he boasted, citing months of testing that the city had done alongside a private manufacturer. That moment of anthropocentric hubris was short-lived. By 2018, one year after a citywide rollout of the new green bins had been completed, widespread complaints began to emerge that they were no match for the *Procyon lotor*. The bins were designed to remain upright and had added security owing to the circular locking mechanism on their lids; nevertheless, they were soon being found on their sides, the locks picked, and their contents torn apart. Initially, city officials chalked it up to a few faulty units and loose screws. Shortly after, videos emerged of raccoons carefully climbing up the bins, sitting on the lids, and reaching out to grab the circular lock handles with one forepaw before delicately threading them back through the mechanism to reveal their delicious guts. No opposable thumbs needed. The city had little to say after that.

The failure of Toronto's raccoon-resistant green bins is evidence of the complementary strategies the animals use to so effectively colonize cities. It starts with their cognitive flexibility, which allows them to unlearn a failed strategy and adopt a new one when their environment shifts. By doing so, urban raccoons exploit a variety of food sources in one area, thereby maintaining small home ranges, which keeps the risks they face to a minimum. These small ranges also increase the size and density of raccoon social networks, which in turn means innovations are efficiently transferred from lone geniuses to tight gazes to the population at large. Toronto's total area covers more than six hundred and twenty square kilometers, far larger than the three-block radius most raccoons that inhabit it will travel. And yet, within just a year or two, raccoons in every one of Toronto's 140-odd neighborhoods had cracked the new bins: The adaptation had traveled through the city's intersecting raccoon social networks like electrical impulses through ganglia. It's an amazingly efficient process of social diffusion, and it explains how the adaptive carnivorans develop innovations and transmit them across urban populations. But it leaves one question unanswered.

How did they open the raccoon-proof lock without opposable thumbs?

■ ■ ■

There's a general rule in biology: Organisms that are more closely connected genetically will be more alike than those that are distantly related. To find the connection between the old oak tree and the young giraffe, we have to go back 1.5 billion years to the moment when plants and animals split from their most recent common ancestor, a single-celled eukaryote. Within the kingdom of Animalia, differences—between the jellyfish and the elephant, the rhino and the rhino beetle—become subtler as you move through each subsequent stratum: phylum, class, order, family, genus, and finally, species, each successive grouping reducing the differences between types. All members of the Carnivora order (including bears, cats, weasels, wolves, pinnipeds like seals, and almost three hundred other creatures) share a few select traits: They eat flesh, have specialized teeth, and walk on four legs. As a carnivoran, we would expect the raccoon would most resemble its closest relatives, like skunks, weasels, and the red panda. And it does—until we get to its forepaws. And that, as biologists have put it, is "the problem posed by the raccoon."

A raccoon's forepaws are highly complex tools almost as protean as an octopus's skin. At one moment they'll look almost exactly like a dog's paws, the same smoothed-over fur covering forward-facing digits designed for lateral movement on four legs. But then, suddenly, the raccoon will lean itself back and hold up its arms, and what seemed like stiff-socketed toes will suddenly spread open into five outstretched, disturbingly human-like fingers. From an evolutionary perspective, raccoons simply shouldn't possess such fine-grained mobility—there are no other carnivorans that have evolved to use their forepaws quite like raccoons do. Oddly, raccoon forepaws more closely recall the hands of their distant relatives in the Primate order than their carnivoran cousins with whom they share the most DNA. And though they don't have the signature primate ability to connect their thumb to their other fingers, they've found hacks to manipulate objects as if they did. The first is through a "scissor grasp," in which they separate their fingers and hold objects between their second and third digits. This allows them to hold and twist objects with a high degree of mobility compared to their peers. Raccoons have also learned to place objects between their two clasped palms and carefully rotate or disassemble them, an action that makes them look as if they're holding their hands together in prayer. Neither of these techniques requires opposable thumbs. What they do require is a fine-tuned way of receiving and interpreting visual information.

Grasping an object is a multistage process: distinguish it from its surroundings, isolate it visually, guide your arm to meet it in space, then contort your hand (or paw) to hold it. This process, called "visual guidance reaching," is a symphony of senses working together in complex ways: The eye has to continually focus back and forth between the object, the hand, and the surrounding space, while the brain must interpret these rapid shifts from distant to proximal visual information. Then the central nervous system has to selectively send impulses along the motor neurons in the spine, arm, and hand to make sure the target is hit. It's an embedded skill among primates and serves as one of the foundations of human supremacy. But among carnivorans, this kind of maneuver is simply too cognitively complex to pull off—with the exception of raccoons.

Mimicking human dexterity, as raccoons have done with their forepaws, would be enough to carve out a niche in cities. But raccoons do us even better when it comes to the power of their touch. Even in the absence of any visual cues or odor signals, raccoons can use their forepaws to locate objects with incredible precision. That's owing to massive clusters of nerve endings in their forepaws along with a huge proportion of the surface area—as much as 87 percent—of their primary somatosensory complex (the section of the brain that receives signals from different body parts) devoted to their paws. For context, in dogs and cats, the same area of the brain makes up closer to 15 percent, while in humans, about 20 percent of this brain region is devoted to receiving information from our hands. But it doesn't stop there. Germans call raccoons *Waschbären* (washing bears) and the French call them *ratons laveurs* (washing rats), owing to their habit of wetting their hands before groping around for objects. Raccoons don't actually wash their food, though. Instead, wetting their forepaws radically amplifies the sensitivity of their touch by loosening the tough skin that covers nerve endings, giving them a clearer tactile picture of what they're searching for. This allows the animals a remarkable ability to "see" with their forepaws, even in total darkness.

Raccoons are believed to have evolved from nocturnal creatures that hunted for crustaceans in swampy waters at the edges of forests, which explains the extreme sensitivity of their paws, the power-up they get when they add water, their fine motor skills, and their habit of relying on touch rather than sight to find food—a must when you're tracking shrimp in a swamp in the dead of night. Of course, these also all serve as useful adaptations to urban living, where raccoons must unlock compost bins and garbage cans before separating out bits of food from inedible trash under the light-polluted darkness of the city.

■　■　■

In Toronto, raccoons have become so ubiquitous that they've moved from nuisance to cult figure to a celebrated part of the city's brand. Toronto's local airline uses a raccoon as its mascot, as does a downtown microbrewery and innumerable coffee shops. Still, after weeks of trying and failing to glimpse raccoons across the city, it takes my encounter in the back of Gordon's van to shake out some sort of metaphysical knot. They are, all of a sudden, everywhere. That same evening, a family—mother and four kits—skitters down the street in front of me, fleeing in the half-hearted way of so many synanthropic animals whose neophobia has faded.

A few days later, a contact at Union Station who I had been haranguing for help in tracking the sources of those pawprints covering the station's skylights, messages me out of nowhere:

> The ceiling of a lunch room just caved in on the 2nd floor from a burst pipe and a raccoon has been spotted living up there!

Later:

> I stand corrected it wasn't water damage the guy had just eaten through a lot of ceiling tiles so the floor is a mess and now there's a trap precariously hanging half in half out of the ceiling and I think they trapped him with rainbow marshmallows and it's super sad and he's still up there

Along with the message are photos: pieces of perforated tile from the drop ceiling lying on the staff lunchroom floor, covered in complicated raccoon-claw etchings; a massive raccoon trap, hovering in the open ceiling, greasy brown fur poking out between its bars; and the raccoon in question, trapped in a wire cage balanced on a stepladder, the beast looking mortified, two identical microwaves side by side on a counter behind him, the walls plastered with posters gently admonishing coworkers to clean up after themselves.

> Update! He has been "rescued" though his exact fate is a little unclear . . . where do these displaced creatures go?

I picture Rob Gordon, or someone like him, taking the animal on a bumpy ride to High Park, Toronto's unofficial home for troubled raccoons.

At home, a few days later, sounds of screaming draw me outside. In the antiseptic blue-white glow of high-efficiency streetlamps, I watch as three fat raccoons wrestle one another up and down a tree trunk, grunting like apneic sleepers and clicking like dinosaurs as they negotiate some low-level territorial conflict in the darkness. Police sirens blare as cruisers speed by. I turn to again look at the raccoons; they stop for a moment and look back at me. In that moment, all four of us are too absorbed by the circular negotiations at hand to care about the world at large. We share this concrete- and tree-lined micro-environment, one ill-fitting puzzle piece among many that make up the city. We stay there for some time, and though their faces are shrouded in darkness, I can feel their eyes fixed on me from above: quietly superior, curious, and dispassionate, and, most of all, unafraid. I guess that's why they call it a gaze.

■ ■ ■

Raccoons are such ubiquitous urban pests that they have, as Rob Gordon put it, become an equalizing force for city-dwelling humans. But that doesn't mean all raccoons are equal. In fact, rural raccoons—which are in no meaningful way wired differently from those in urban spaces—struggle to challenge their urban cousins on cognitive flexibility, problem-solving, and persistence. The extent of this gap was revealed in a series of experiments conducted in backyards and outdoor spaces across Toronto and rural parts of Ontario. Knowing how food-motivated raccoons are, scientists created two puzzles: a hanging bucket suspended upside down about two feet off the ground from a rope tied to surrounding trees, and a standard plastic garbage bin with a bungee cord affixed on top to deter removal of the lid. Both were loaded with food. After compiling eight hundred hours of video taken over two years, and then removing sections of tape featuring black bears, coyotes, and domesticated cats, scientists were left with footage of twenty-two rural and twenty-two urban raccoons.

Both the rural and urban groups made quick work of the hanging bucket, which required only that the raccoons duck under it and then reach up to access the food. But when it came to the second puzzle (the bungeed garbage-can lid), stark differences between city and country emerged. More than three-quarters of the urban raccoons absolutely savaged the garbage cans, tearing off the bungee cord with their dexterous fingers and gorging themselves with abandon on the food inside. And while not all the urban raccoons succeeded at first, they also refused to give up. Instead, they ran through multiple strategies, like trying to sever the bungee cord, break the lid, or flip the bin over, until they finally succeeded in shaking loose the contents and feasting. But of the twenty-two rural raccoons, not a single one of them could crack a bungeed garbage can. Many quickly resigned themselves to defeat, while others did so after trying the same strategy over and over again, evidently expecting a different outcome. "These data," wrote the scientists, "support the tantalizing possibility that the anthropogenic selection is at work, with our cities—and human behavior—selecting for particular cognitive abilities in raccoons." In other words, while the raccoons of the countryside had stayed hungry, the city had attracted animals gifted enough to keep their bellies full.

This all might seem a little overwrought. After all, we're talking about a bungee cord and a garbage can. But the absolute failure of rural raccoons to outcompete their urban counterparts is shocking, because there are no genetic differences between the two. And this is where these findings start eroding some long-held assumptions about animal biology and the evolution of intelligence.

Evolutionary biologists have developed four theories about how intelligence evolves. The first, known as "social complexity," suggests that intelligence is selected over time because possessing it maximizes social agility—that is, the smarter the animal, the more safety (and power) it holds within its social group. The second theory, known as "cultural intelligence," builds on the idea of smarts bringing social power but proposes that the key benefit of intelligence isn't social inclusion but the learned skills that go along with being part of the group. The smarter you are, the better positioned you are to learn and share new things. Then there is the "cognitive buffer" hypothesis, which suggests that as animals move into new and unpredictable environments, only those with greater intelligence will be able to survive the chaos, which over time will cause only the more intelligent individuals to be selected for survival. Finally, "ecological intelligence" proposes that acquiring food through a mix of strategies—foraging, hunting, and using tools, and doing any of these within large social groups—rewards a truly multifaceted intelligence while minimizing the costs of failure. In the case of early humans, if a hunter tried a new technique but failed, they would still be provided with food by others in their social group. That meant that risk-taking could be rewarded without causing more daring individuals to be punished when they came up short.

All of these theories plausibly explain human evolution. They're also predicated on extremely limited data, because we don't have much information about how humans evolved prior to our emergence as the world's apex predator. That's created a serious barrier to testing these theories because the moments that spurred human intelligence to reach such astounding heights (isn't it incredible that I can write these words and you can understand them?) are all in the distant past, blurry ghosts predating the Neolithic Revolution. The only way to know which of them is correct is to witness intelligence evolving in another species.

Theories explaining the evolution of intelligence have focused almost exclusively on humans and other primates. In recent years, though, the movement of animals—especially carnivorans—into human-made habitats has opened up the possibility that we could be witnessing in real time the rise of a new kind of intelligence. Cue the urban raccoon, the garbage can, the bungee cord.

With a flood of new studies showing that urban animals like raccoons are able to think differently from their rural brethren, it seems increasingly clear that city living is a mind expander, consistent with the cognitive buffer theory. One hypothesis is that raccoons have long had these cognitive traits embedded in their genetic code but haven't needed them until city environments forced their expression. (How can you know you're a great dancer if you've never heard music before?) Another is that cities, with their complex dangers, irrational spatial organization, and intricate puzzles are selecting for smarter raccoons, those that see a patchwork of disconnected microenvironments as a challenge to be conquered rather than a hellscape to be feared. Wherever this may ultimately lead, it's evident that cities, far beyond simply harboring animals, are changing the way they think.

Raccoons are impressive urban adapters, but they're also relatively unique among synanthropes in having an outsized capacity for cognitive learning paired with a rapidly learned fearlessness. So how is it that more simpleminded species, and those that retain a healthy neophobia, are drawn toward our homes? Sometimes it's not enough to look at animals themselves. Instead, we need to look at the evolution of cities—and especially what happens when they start to fall apart— to understand how ecological niches can attract and transform all kinds of creatures, including those profoundly unlike ourselves.

Chapter 2

The Tower

Remember when
we both first felt
the world is sound,
the world is sound?

—Hot Chip, "Motion Sickness"

Cities Are Islands of Evolution. Cities are so ecologically different from the areas surrounding them that they become islands of animal activity and adaptation. And when urban areas sink into decay, new ecological niches open up that can amplify how cities both attract and isolate animals. It's in these precarious urban environments where some of the most remarkable adaptations are taking place, as synanthropes reclaim spaces that we have all but left for dead. In some cases, these adaptations are so extreme that they have the power to make an urban animal unrecognizable to its own kind.

■ ■ ■

It was one of the first parties that Joyce Hwang attended after moving to Buffalo, New York, from Philadelphia in 2005. And it was a big deal: This city, rusty and gothic, was her home now. If all went well, it would be for the rest of her life. She was a newly minted tenure-track professor at the University at Buffalo with a bright future in architecture. Still, designing buildings can only take you so far in life. She also needed to find her place in the city.

When the party conversation shifted to Buffalo survival tips, Hwang was all ears. Amid recommendations about where to shop and what streets to avoid, the assembled group all agreed on one thing: Hwang needed a tennis racket. "We have a ton of bats here," someone said, the group nodding in agreement, "and you can only kill them with a racket." Hwang was aghast but tried to conceal her shock. *Why would I swing a tennis racket at a bat?* she thought. *It doesn't make any sense.*

The bloodlust was disturbing. As an architect, Hwang was trained to design buildings that make communities feel more welcoming. Unbeknownst to those at the party, Hwang wasn't just designing for humans. By the time she arrived in Buffalo, she had already spent years thinking about how architecture could evolve to better allow human communities to cohabitate with animals. Back in the early 2000s, though, nobody in her field really got it. Her professors demanded that she explain her rationale for designing buildings with animals in mind. When she tried to explain, they responded brusquely. "What are you talking about?" they would ask her. "This is not architectural work." While Hwang understood that she was pushing architecture in a new direction, she still figured there was enough room in the discipline to experiment with new forms. But academia can be a brutal, uncompromising place, and Hwang still worried that she would never find a real place among her peers, even after she got the coveted tenure-track job with the University at Buffalo.

Partially to understand her new home and partially to clear her head, Hwang began to run, traversing Buffalo neighborhoods dotted with abandoned houses, outdated infrastructure, and vacant lots, ugly and overgrown, that had been reclaimed by nature. In 2005, the year she arrived, the city had just emerged from one of its hardest decades. Rising inflation and interest rates had sparked a rash of foreclosures, which had been steadily ramping up since the 1980s. Upstate New York had fared worse than elsewhere, and the Niagara-Buffalo central core was a blinking red warning light of what can happen when a city begins to buckle in on itself. In the 1990s, Buffalo experienced a four-fold increase in annual foreclosures, from about two hundred homes seized by lenders in 1990 to more than eight hundred by the end of the decade. During that time, the city's employment rate had remained low and static, though the United States' national rate had risen by more than 20 percent, effectively leaving the city behind. Buffalo was also mired in stagnation, having relied on steel production for so long that it didn't have a coherent plan for a deindustrialized future. Though the industry had ceased to operate in the region by 1983, taking seventy thousand jobs with it, little had taken its place: There was no dot-com bubble, no transition to green tech, no secondary manufacturing industries, and few jobs to go around. Unemployment was at 13 percent, and the city seemed to be sleepwalking into a bleak future.

The neighborhood that Hwang settled in, Elmwood Village, was one of Buffalo's foreclosure hot spots, and hundreds of buildings around her apartment were in receivership. By 2005, there were twenty thousand vacant homes in the city, painful monuments to urban failure and fragility. Perhaps "vacant" is the wrong word, because Buffalo had over time become home to nine different synanthropic bat species, which had found a wealth of roosts and hibernacula hidden within its long-decaying infrastructure, much of it a relic of the city's boom time in the early twentieth century. Back then, the local economy was powered by a wartime need for steel, chemicals, aircraft, trucks, and ammunition, which Buffalo's heavy industry produced at scale. The profits from those industries had been transmuted back into stately buildings cast in a variety of styles, including Federal and Gothic Revival, their pillars and painted cornices giving the city an eerie, fairy-tale quality that lingers to this day. More than fifty years later and the result was a dense network of stone buildings, never too hot or cold, many left vacant or undisturbed for years at a time—in short, a synanthrope's dream. Better yet, from a bat's perspective, Buffalo is situated in a wet and open region that spans lakes, a river, and innumerable marshlands filled to the brim with a predictable supply of mosquitoes, dragonflies, and other protein-rich invertebrates. For bats, it's as close to perfect as a habitat can get. For the type of human who wields a tennis racket as a weapon, living in a bat utopia is a nightmare.

Hwang quickly realized that this unassuming and deceptively static town was teeming with animal life. She passed raccoons that had set up shop in a dilapidated building down the block from her place and came home to some living in the tree in her backyard. She saw Canada geese and black skimmers migrating above her along a path that spanned the hundreds of miles between Lake Erie, Niagara Falls, and nearby Buffalo River. She tracked deer loping along underused railroad tracks, their coloring ferrous and bright. But what captivated her most were the animals she only rarely saw. Everywhere, Hwang felt the presence of bats, minuscule and out of reach, flooding the sky at dusk to feed on the great insect biomass of Erie County.

It didn't take long for Hwang to understand just how right the group at the party had been. Buffalo is uncommonly inundated with bats, which have found a niche in the city through a confluence of environmental, economic, and architectural factors. But instead of picking up a tennis racket, she started playing with paper, sketching out what kinds of homes she might be able to build for the animals that surrounded her.

▪ ▪ ▪

Murder by blunt force is most efficiently done with solid mass, the tension of the tightly packed atoms of a piece of wood, say, increasing the energy that can be applied against a hunk of flesh and bone. It's the rare animal, then, that is more susceptible to a tightly bound bundle of crosshatched strings than a hard surface. But the bat is a rare animal indeed.

Not just rare, in fact, but eerie. Here is an animal with the patina of kinship with our kind—warm blood, oxygen-pumping lungs, downy fur—that has perverted those shared attributes to become something grim and grotesque: a fluttering, nocturnal, ostensibly half-blind predator that navigates its world by seemingly supernatural means. It spends its days upside down and its nights hunting in the dark, the topography of our homes clearly illuminated by senses we cannot perceive. We fear bats. We fear their numbers as they swarm, which in some colonies reach tens of millions. We fear their strange skittish fluidity as they trace chaotic patterns in the night sky. We fear that we can't see them but they can see us. What they do and what they are feel like an assault on what we know in our hearts makes a mammal, a cousin, kin.

Once you peer into the class of Mammalia, though, the picture becomes as inverted as bats themselves. Bats, order Chiroptera ("hand wing" in Greek), first evolved from their most recent common ancestor during the K-Pg boundary, a period 70 million years ago that marked the transition between the Cretaceous and the Paleogene, when Earth experienced one of its greatest mass extinctions and the end of the dinosaur age. By contrast, our genus, *Homo*, which includes *Homo sapiens* as well as all extinct humans, emerged only 5 million years ago. When our first human ancestors finally arrived on Earth, early chiropterans had already spent about 45 million years adapting to this world, perfecting the tools they've used to colonize every global hemisphere, and giving them a far more ancient claim on the land (and sky) than the one we hold.

How they did that, though, is shrouded in mystery. Bats have left the poorest fossil record among all four-limbed animals, estimated at about 15 percent completeness, likely due to the fragility of their bones. All that scientists have been able to piece together from this incomplete record is that the proto-bat ancestor was likely a small insectivore with four legs and paws. But how something that resembled a shrew, wingless and rodent-like, could then assemble itself into 1,483 (and counting) unique species, all with the ability to tumble down into flight, eludes us to this day. And while that mystery remains unsolved, our persistent notion of what constitutes a "normal" mammal—earthbound and upright—is simply wrong, because almost a quarter of them are bats. That means that, among the 6,400 or so extant mammalian species, the bat is among the closest there is to a standard-bearer.

There are bats, like the hammer-headed fruit bat, that resemble horses with wings and make sounds like bouncing cartoon pogo sticks. There are others, like the male of the dayak fruit bat, that lactate to provide fresh milk to their young. To cling to the smooth cylindrical leaves of banana plants, the Spix's disk-winged bat has evolved suction cups instead of claws; it looks like a sticky cotton ball with wings. Then there is the Mexican free-tailed bat, its ornate pointed ears resembling the high crown of a medieval king, its long ergonomic tail allowing it to reach flight speeds of 100 miles per hour, making it the fastest mammal on Earth. The dazzling variation among chiropterans has meant that bats have mastered practically every habitat there is (Antarctica being the only exception). They are the world's most adaptable mammalian order, and it isn't even close.

They're also, by some measures, the world's most successful urban vertebrate. Across the globe, about eighty different species of bats have become urban dwellers, dwarfing the diversity of any other vertebrate classes. Human-built environments, a recent wrinkle in their long evolution, have become just another domain that bats have conquered. This ability to adapt to many different environments is called "niche generalism," and it's a trait common among synanthropes. Many bats are generalists, with a broad diet and the ability to expand their range into new territories if necessary (that's in contrast with specialists, which are species that depend on highly specific habitats and food sources to survive). And while there are numerous reasons why many bats are such successful generalists, one simple reason stands out: Theirs is the only order among Mammalia to have achieved powered flight (with apologies to flying squirrels, who do the best they can with the tools they've been given). This clears several hurdles that urban spaces present to animals unlucky enough to be permanently grounded. Raccoons have had to rely on cognitive flexibility to navigate the ill-fitting puzzle of urban spaces. Coyotes have had to overcome their latent neophobia (a product of evolving as scavengers alongside wolf packs) and become increasingly nocturnal to survive in cityscapes. And rats have evolved at lightning speed to weather the barrage of poisons that humans have used to try to kill them. These adaptations represent major cognitive, behavioral, and genetic leaps. And then there are bats, which haven't had to twist their bodies and behaviors into strange new forms. Instead of wild leaps in cognition or advantageous genomic mutations, they just brought their wings.

■　■　■

At a global scale, scientists have found that cities with older residential areas, nearby rivers and lakes, and lots of parks are a magnetic draw for bats, making Buffalo an ideal habitat. The city is also home to about 275,000 residents—fewer than in 1895 and less than half of its peak population of 580,000 in 1950—leaving lots of room for bats to roost undisturbed in abandoned buildings. In the city's downtown core alone, more than 5 million square feet of office space, mostly in older, no-frills building blocks, lie empty. With an abundance of insects from nearby Niagara River, and a blessed absence of predators (except for racket-wielding hominids), bats have pushed Buffalo's slogan— the City of Good Neighbors—to its limit.

Beyond these amenities, Buffalo and every other city provides another benefit rarely found in nature: a continuous heat source. And for bats, which are tiny things, with a thin layer of skin stretched tautly across their wings, heat is more sacred than for most other hot-blooded mammals. Picture the scene from high above: It's nighttime, and the pleasing landscape of seamlessly connected woodland, snaking rivers, marshland, and pastures is shrouded in darkness. It's autumn too, late October, and it's getting cold. Normally, this is the time when cave-dwelling bats return to their hibernacula and enter torpor, a sleep so deep it mimics death, to survive the harshness of winter, and when migratory bats leave Canada and the northern United States for the healing warmth of southern climates. Torpor and migration are two strategies bats use to stave off winter, as if they are trying to warp time itself.

But cities warp time so that bats don't have to. They do it through the "urban heat island" effect, which refers to the unique way that heat is produced and trapped in cityscapes, creating weather systems that become self-contained and, consequently, much hotter than those around them. In some cases, these effects are extreme: Tokyo, Kuwait, Phoenix, and Las Vegas can become close to 30 degrees Fahrenheit warmer than surrounding rural areas. But it's not just the heat that attracts bats: It's how static urban environments are. And that's how cities are completely rewriting the life cycle of an animal whose behaviors are deeply linked to seasonal change.

In winter, there are so few insects that hunting results in a net loss of energy. Worse, the arrival of subzero temperatures robs the animals of liquid water, which is tantamount to a death sentence. Bats that enter torpor are hardwired to respond to these cues by slowing down their metabolism (some bats go from eight hundred heartbeats per minute to as few as eight), dropping their body temperature, and falling into a dreamless void. When bats descend on urban heat islands, where temperatures are kept artificially high even as autumn turns to winter, the triggers that hardwire them for torpor—cold and famine— either are delayed or sometimes never occur at all. This allows bats the freedom to be more active throughout the year to hunt, quench their thirst, and mate; it's endless summer in the city. It's also one of the reasons why some synanthropic bat species are much more abundant in city centers compared to surrounding areas. In Buffalo, that includes the little brown bat, a.k.a. *Myotis lucifugus* ("mouse-eared light avoider"), a chiropteran as long as a finger, which is among the most plentiful in upstate New York. One colony of breeding females has been active in a nondescript barn in the town of Westfield, about an hour south of Buffalo, since at least the Civil War. Over the 165 years since Americans took up arms against one another, Westfield has expanded and so has the impact of its urban climate zone on temperature regulation. That warmer weather, a key feature of urban living, has accelerated juvenile bat development, so bat pups in the area are growing faster and learning more quickly, like city kids forced to toughen up and get streetwise in a hurry.

All cities produce their own local atmosphere. It's known as the "urban boundary layer," and it mixes with a larger planetary boundary layer to create a climate system that sits over a city like a dome, cutting it off from the surrounding world. Within this dome lies the urban canopy, a rough and temperamental layer of air subject to the vagaries of the buildings, monuments, waterways, and (if we're lucky) trees and plants that make up our communities. Some of us think of condo buildings and office towers as blights dragging the city down or as necessary evils; atmospheric scientists are less interested in how we use buildings than in how they force air and heat to move around urban environments. To them, buildings make up a canopy akin to those found in forests or jungles: the highest-elevated point in an ecosystem that harbors life, and has an essential role in managing how and how much light, air, water, and pollution flow in and out of the habitat below.

In a forest, the canopy also has an overstory. This is a matrix of tree branches, crosshatched and overlapping, gnarled and directed in their growth by nature's meta-intelligence, that acts as a living membrane, keeping the forest shaded from atmospheric extremes. In the urban canopy, there is no such overstory and no such regulation. Buildings are arrested at right angles, their walls dropping so steeply and with such precision that ecologists call them "urban cliff faces" and refer to a cluster of them as a "street canyon." The canopy they create is nothing like the diaphanous living roofs of forests and jungles. It is just wide open, unable to regulate the exchange of atmosphere from above and below, and instead allowing sunlight, rain, dust, and heat unimpeded access to the urban environment under the dome. Under the urban canopy, the gentle rhythms that serve as cues for animals to move through their life cycles are gone. In their place is a riot of sensory signals so loud and chaotic as to be indecipherable.

With so little cover, you might expect that urban atmospheres would be more volatile than those in natural spaces. And yet, not only are cities much hotter than the areas that surround them, their temperatures also tend to fluctuate far less, putting them in what urban ecologists call "disequilibrium" with surrounding areas. To understand why, all you need to do is go outside on a hot day and lie down in the middle of the closest road. Immediately, you'll feel the heat trapped in the tarmac below, searing to the touch. Get up off the road and go lie down in a patch of grass nearby, and you'll be struck by the relative coolness of the blades tickling your neck and the backs of your arms. The materials we use for construction—concrete, but also plastics, steel, and glass—are mostly excellent at storing and conducting heat. When the sun rises, urban materials begin trapping heat, absorbing it and storing it all day across the cityscape. As the sun dips below the horizon, the heat stored in roofs, walls, and roads is drawn out into the cooling night, where it diffuses across the darkness of the urban dome, keeping the urban heat island warm. At dawn, as the grease from a twenty-four-hour diner mingles with the vestiges of last night's synthetics, and you walk home within an urban canyon with a strange kind of temporary balance, the cycle starts anew: hot days absorbed into the very fabric of the city, mild nights as the energy is released, all of it under the open urban canopy. We look up at the light-polluted sky as we meander through cities, while bats, seeking the warmth of urban heat islands, descend upon them from above, the city exerting an invisible force, drawing the creatures down into our homes, promising them comfort.

■ ■ ■

The bloodlust and raw emotion that bats elicit in people is exactly why Hwang was attracted to the creatures. So, against the advice of her colleagues and mentors who considered architecture a human game, Hwang set herself the task of not only communing with Buffalo's bats but also building them homes that could be celebrated by all. I was familiar with Hwang's work before we met up, having experienced a sculpture called *Multispecies Lounge,* designed by her and the architect Nerea Feliz, which is an assemblage of thick wooden planks standing upright, topped with roosts that would be ideal nesting spots for birds. The wood was perforated throughout with different-sized holes, which at first seem like a random set of markings until you realize they're designed as places for synanthropic solitary bees to lay their eggs. Set horizontally are benches to accommodate *Homo sapiens,* while underneath, small mounds of rocks wrapped with chicken wire provide lizards with a warm hiding place and protection from the elements. Sitting on one of the benches on a sunny day, I vibrated with anticipation, hoping that one of the other animals the lounge was meant to host—bird, bee, lizard—would come and visit, before realizing how rare it was that I ever even thought about the paths wild creatures took through the city. It was a lesson in how Hwang's design could profoundly alter my mindset toward urban nature.

I meet up with Hwang on a warm August day shortly after her return from a six-week artist residency in Umbria, Italy. Bunking with a dozen or so other artists and writers in a fifteenth-century Renaissance castle, Hwang had been invited to participate in a collaborative art-making project. Instead, and true to form, she found herself living nose-to-snout with a castle full of bats. They hung along door lintels and under the desk in her studio. Slide back a painting of some aristocrat in one of the rooms, and there, in a perfectly concealed crevice, was a sleeping bat, deep in torpor and grousing quietly about being disturbed. Hwang, who has had lots of experience chasing the animals through Buffalo and building them homes, became the group's unofficial bat expert.

As we head to her office on the north campus of the University at Buffalo, Hwang points to the building next door, Georgian Revival style, with an off-design white-painted steeple conspicuously planted on top. "Hayes Hall," she says as she searches for her keys. "That was where my office was when I first started here. It's a former insane asylum," she adds with a bright smile. I squint, and though it's the middle of the day, I can see it: the black silhouette of the building in the dead of night, like it's got its back up, lightning bolts flaring in the distance, maybe some maniacal howls, the whole scene right out of a Batman movie. Built in 1874 (and known back then as the Insane Department of the Erie County Almshouse), Hayes Hall hadn't been seriously renovated in over half a century when Hwang first started working there. That meant, she says, there were bats everywhere. A few weeks after starting her job, Hwang rushed to catch the elevator to her office, only to be confronted with a handwritten sign reading "Avoid—Bats in Elevator." The steeple was full of them, she was told; bats would slip into the tower and into the crumbling walls to find roosts, which were the right mix of warm, damp, and undisturbed. "Back then," Hwang says as she scans her key card to enter her new office space, which is bright white, temperature and humidity controlled, and has motion-activated lights leading us forward like a trail of candy, "the faculty mailboxes were these thin wooden slots that stretched far back into the wall, and people would always be finding bats with their mail." Eventually, after weeks of bat infestations, Hwang arrived one day to find that the handwritten warning sign at the elevator had been replaced with a printed one, a formal acknowledgment that the bats were there to stay.

We enter an antechamber within Hwang's new office building, where she was moved a few years back. The building is empty, a perfect roosting site for bats before it was renovated to meet every conceivable standard for low emissions and green design. What I see of it strikes me as slightly sinister and altogether antiseptic. When we enter Hwang's office, by contrast, it feels comfortably full, books and papers in neat piles, the classic picture of an academic at work. As an architect, she's sympathetic to notions of sustainability and understands the need for the renovations, though she admits that she misses the old building: There was a wildness to the space, she says; it felt rough-and-tumble. There were also way more bats all over the campus. As she sits at her desk, her cell phone suddenly lights up. Hwang frowns slightly as she reads the message. "This happens a lot," she explains. "Every time someone sees a bat I get a text."

Scattered amid the comforting disorder of her office are three-dimensional architectural renderings that read as art pieces rather than utilitarian structures. Hwang calls them "speculative models," early drafts of principles that she might use for future work. One maquette consists of a two-sided bristol-board pyramid sprouting a stack of about a dozen chopstick-sized wooden batons placed at convoluted angles. Another resembles a mock-up of an ancient, walled-in temple, the inner chamber set with a series of shelves, each one shorter than the last; in the outer chamber, a far larger space, a horizontal series of thin wooden planks of uneven lengths have been placed one on top of the other. I ask Hwang what the shelves are for, and she smiles, takes the model, and flips it ninety degrees, turning it into a long vertical space. It's called *Habitat Wall,* she explains, a home for birds and bats, one of many that she's developed. What I imagined were floors are its walls, while the shelves are actually vertical slats for the bats to cling to while they sleep.

As Hwang takes me through her work, it's evident that she is building out a new discipline: the design of structures that appeal to the human eye but serve a multitude of species. There's *Bat Cloud*, consisting of narrow tunnels placed in trees, below which hang about a dozen plastic baskets and planters looking like wispy forms emanating from the night. Hwang and her team installed *Bat Cloud* in the Tifft Nature Preserve, a 264-acre green space that adjoins downtown Buffalo, and the animals quickly moved in—that is, until the fragile edifice started to disintegrate (it's an occupational hazard in her line of work). And then there's *Bat Tower*: a wooden sculpture that looks like a miniature wizard's castle made of three-dimensional triangular shapes stacked on top of one another. Looking at it, you might think it's an inspired piece of land art, but it's actually a cleverly devised home for bats, complete with a hollow interior, landing pads, and a garden of wild herbs growing at its base to attract insects for its residents to feed on.

I ask Hwang about her first experience with bats and she becomes wistful. "It was maybe in someone's backyard," she says. "It's at a certain moment in time and there's just enough light in the sky, but it's still dusk, and I can see them flying."

"And how did it make you feel?" I ask.

Hwang pauses and wrinkles her nose in thought. "Well, it's beautiful," she says after a beat. "And now that you're asking these questions, I can't even pinpoint a moment when I thought, *Oh, that's the first time I saw a bat*. It's almost like saying, 'When did you first see a bird?'" What Hwang remembers vividly, though, is the feeling of their absence in places where she used to see them regularly. It's an experience she's had a few times in different places, including Buffalo, the Spanish countryside, and her eco-friendly and anodyne new office. "Bats used to feel atmospheric, like they were just part of what happened. And it felt really strange when I didn't see them." She never wanted to feel that again. So she decided to build them homes.

All of Hwang's projects confront a fundamental reality of conventional architecture. "The fact is," she says, "every building we design takes away land from every other species on Earth." Even supposedly "green" initiatives are way too human-centric, she argues. "Creating a green roof"—an increasingly popular design choice for skyrises that entails covering a roof in vegetation—"can be really problematic. If it's too high off the ground," she points out, "it'll attract birds, but the babies will fall and die when they're learning how to fly." For Hwang, the only ethical way to be an architect is to design specifically for non-human species rather than making ecological concerns a tangential part of the process. As she's gone down that path, she's been forced to leave traditional architecture behind. It can be a difficult journey at times, as Hwang has tasked herself with developing a whole new approach to communicating her ideas. Meanwhile, the winged residents of her sculptures are striving headlong toward that same endeavor: finding ways to animate urban spaces by creating a new language.

■ ■ ■

Bats have colonized our cities from the sky, making them among the world's most successful synanthrope. But on land, one of the most successful vertebrate synanthropes is also among the first to live among us. The red fox, *Vulpes vulpes,* is a slim and canny mesopredator with a home range stretching from the Arctic Circle to southern North America, Europe, North Africa, India, and Japan. That geographic range reflects its innate skill in exploiting all kinds of environments—including urban ones.

Dawn Scott is a leading expert on urban red foxes in the United Kingdom, a job she took after motherhood made traveling to exotic locales to study wild animals effectively impossible. After studying human-elephant conflict in Zambia, human-tiger conflict in Indonesia, and the impact of logging on biodiversity in Madagascar, the drama of red foxes padding through the cobblestoned streets of London carried its own kind of thrill; the first time she saw one in the English capital, she says, was "mind-blowing." Scott explains that foxes, just like raccoons, are among a class of mammalian mesopredators that are primed to exploit cities but have achieved a global range unrivaled by any other carnivore. "They're quite agile," she explains, "so they can navigate in a three-dimensional space. Unlike some of the other urban animals, which can't get into gardens, foxes can jump over fences. Plus being intelligent enough to find food and adaptable enough to eat anything," says Scott, opens up a range of possibilities. Perhaps most important, she adds, is "not being too big to be scary."

Though evidence of fox synanthropy dates back twenty-seven-thousand years ago to the Pavlovian culture, who wore amulets of fox teeth, it was only in the 1930s that Britain's foxes began to move into urban centers. One theory is that World War II caused rural country estates to cease employing gamekeepers; without population controls at the barrel of their guns, fox numbers exploded. For most animals this would lead to greater population density, but foxes self-regulate their population, allowing only about thirty individuals per square mile, forcing the overflow to move out of the country and into cities. Another theory posits that the expansion of suburban communities trapped foxes in city environments, to which they then became accustomed, and from which they could no longer leave.

Scott has become something of a celebrity in the UK, regularly appearing on the radio to talk people through their fox problems, of which there are many; recently, she was a regular on a BBC Four program called *The British Garden: Life and Death on Your Lawn*. And while she's passionate about the animals, she's heard from lots of people who find them annoying. I ask her why and she laughs. "Okay, where should I start? They dig up gardens; there's defecation; when the cubs are born, they can cause chaos—it's like having puppies in your garden. They'll dig and explore and chew everything. The latest is the fact that they scream while breeding at night." I laugh. "A couple years ago, they were chewing the brake cables of cars," she adds, which caused crashes after a number of people's brakes failed. "So, yeah," says Scott, "they get into trouble." Trouble, for an urban fox, includes a grab bag of misdeeds that have made headlines: biting the hand of a four-week-old baby while dragging it from its cot; going on a cat-killing spree across London that resulted in four hundred feline deaths (and was long blamed on a human serial killer); and, famously, sparring with Larry, the resident mouser of 10 Downing Street, the residence of the UK's prime minister (the fox lost). For all their publicity, though, fox attacks are exceedingly rare: between 2010 and 2019, there were only twenty-three reports of foxes biting or scratching people in the UK (this tally does not include an incident when a mother discovered a fox licking the face of her sleeping six-year-old daughter), or an average of two or three per year across a country of 69 million people.

Scott has done comparative work across the UK and found that red foxes seem to adapt to cities regardless of their urban design. Bournemouth, for instance, a southern seaside idyll made up of miles of pastoral green space, is full of foxes, which have found in its plentiful gardens lots of prey and places to hide. So too, though, is London— a city of close to 9 million people living across a highly fractured landscape replete with fences, roads, rail lines, and apartment blocks that should be the worst nightmare of an animal whose home range is wide-open plains and sprawling forests. And yet, there are ten to fifteen thousand foxes living in London, a number that has been stable for decades.

Much of the research Scott has done involves collaring foxes and tracking them over years, which has given her unique insight into their movements and adaptation to cities. "At one point," she says, "we had twenty foxes with trackers in different parts of the country. But, you know, their ranges are incredibly small." Still, with dozens of collared foxes crisscrossing the country, Scott is constantly on call, with fox movements sent to her automatically by text message. If a tracking collar suddenly turns off or a fox ends up somewhere it shouldn't be,

Scott has to find it. That means she's used to driving around at all hours of the night to rescue foxes that have found themselves in bad situations, following a flood of text-messaged directions blowing up her phone. A lot of the time, the problem stems from a fox that has trespassed onto someone's property, which presents unique problems for which her field research in wide-open African plains did not prepare her. "The biggest challenge is that you're working at night in a city—which is quite interesting—but you're crawling around people's gardens," she says. "It's been quite dodgy. We've had people come up to us with samurai swords in the middle of the night, or drunk people."

"Wait," I say. "Stop—you were threatened with a samurai sword?"

"Well, one of my students was," Scott clarifies, "in the middle of the night when he was radio tracking badgers." That's just the way it goes, she explains. The movement of urban animals doesn't align at all with the way humans map out cities. "You know, we've found foxes in heroin dens and we'd have to go in and locate them. We've had people call the police on us. And so you've just got to be aware of what you do and where you're going. You're in all the places that you really shouldn't be at three o'clock in the morning." It's a good reminder that no amount of time chasing elephants can prepare you for the strangeness of urban ecology.

▪ ▪ ▪

Bats, reigning down from the sky, evade the endless set of puzzles that bedevil earthbound synanthropes. Bats haven't had to learn how to overcome fences, barking dogs, or the assault of traffic. Instead, they approach city architecture in three dimensions, scanning high above the fracas to select treetops and church steeples, gutters and billboards, dilapidated homes, manicured parks, and suburban spaces to serve as their homes. But if all it took to conquer a city was wings, we'd be confronted with birds everywhere we looked. Rookeries of albatrosses silently gliding amid the skyscrapers, never landing. Yellow charms of goldfinches as thick as clouds heckling us as we step outside, car fob in hand. But this is not the case. Bats have had to do more than just fly.

Both their success in colonizing human spaces and their weakness against rackets come from their reliance on sound rather than sight to navigate the unpredictable topography of city spaces. The technique bats use, echolocation, involves sending out ultrasonic audio signals (those above the human auditory range) and listening for how quickly those signals echo back toward them. Echolocation, though, doesn't simply mean swapping light for sound. When we open our eyes, we're passively absorbing light beams in a steady stream through our retinas, which makes our experience of the world continuous. Light at night is rare in nature, but so too are predators, and bats have found a way to unlock the darkness to survive. Because they rely overwhelmingly on the echoes from short bursts of sound that they actively produce, bats paint their world through flashes—frozen images jittery in space, shifting with their movements, a world pieced together from fragments to make a whole, like a strobe.

In practical terms, echolocation allows bats to parcel out the world around them into two broad categories: objects and open space. When a bat call flies through space unimpeded, there is no echo. To a bat, this means a void, and the void means safety: an open path to fly through. But when a bat call bounces back as an echo—and bat species make between 5 and 190 vocalizations per second—the animal's specialized sensory organs are able to make fine distinctions to create a four-dimensional portrait of whatever object the call bounced off. The farther away the object is from the bat, the longer the echo takes to return. If the echo returns as a faint whisper, the object is small, reflecting only a fraction of the clicks and squeaks sent out by the bat. If the right ear hears the echo before the left ear, this places the object in a specific location along a horizontal grid. If the echo is pitched higher than the original call, then the object is traveling toward the bat, the sound waves clustering closer together, causing a Doppler effect; if the pitch is lower, then the bat must speed toward its escaping prey.

There's a human tendency to assume that sight is the emperor of all the senses, a more finely tuned way of understanding the world with maximum gradient and a level of nuance unmatched by any other information source. We are phototaxic, a species instinctively affected by light, drawn toward it by some hard-coded neurology that lies so deep within us that we could not ignore it if we tried, while our planet orbits a giant ball of light at the center of our solar system. Compare that to sound, which doesn't rise with the sun or ebb with the stars.

When you try to understand what sound is, exactly, you get perilously close to the absurd. The definition that the American National Standards Institute provides in *Acoustical Terminology* guidelines is practically poetry, and best read as such:

> *Sound is*
> *oscillation in pressure, stress, particle displacement,*
> *particle velocity, etc.,*
> *propagated in a medium with internal forces*
> *(e.g., elastic or viscous) or*
> *the superposition of such*
> *propagated oscillation*

Later, the guidelines veer close to self-parody when they state that sound is also "auditory sensation," evoked by the substance described in that poem above. To clear up any confusion, the guidelines later state that "[n]ot all sounds evoke an auditory sensation. . . . Not all auditory sensations are evoked by sound." Put another way, not all sounds create sound. Compare this tortured abstraction with the definition that the institute gives for light: "Light refers to the portion of the electromagnetic spectrum that is visible to the human eye." Simple.

Understated. Reverent. Light is so obviously the thing it is that its definition doesn't have to do backflips to assure us that it is unmistakably, in and of itself, the most important substance in the universe.

Bats can see, by the way, having typically mammalian eyeballs. Just like us, though, they cannot see at night. But bats cast their own version of light wherever they go, squeezing their larynx and absorbing their own antic voices back as if celebrating their own songs, the tender and intricate folds of their ears, cheeks, and noses interpreting their echoes with an unimaginable level of granularity. In leaf-nosed bats, the most diverse bat family in the world, it is as if their entire faces have opened up like hand-cranked mechanical toys to reveal the inner workings of their minds: folds within folds burst forth in a fleshy baroque, allowing them to better absorb the pulses returning to their bodies, to commune more perfectly with the world that they sing into existence as they fly.

This singing gets complicated when bats enter cities, where incidental noise has become so ubiquitous that most of us accept it as part of the urban experience. Bat biologists refer to cities as "urban acoustic environments" in recognition of how deeply human-engineered spaces have warped the way sound moves. Sound pollution is everywhere: the rumble of cars, the heavy squeal of iron wheels on railroad tracks, the sharp decompression of air brakes as semis roll off the highway and are forced to a stop, wheezing, at construction zones where jackhammers emit sonic shock waves as they break apart hard concrete. When they discharge electricity, high-voltage power lines emit a crackling hum that sounds like barely contained Old Testament anger. A rush of human voices sounds like droplets enveloped within a liquid until a sudden shift in volume and timbre elevates a lone voice above the fray. And that's just what *we* hear. Still, it all feels right. As annoying as the squeak of a subway car is as it takes a turn along old rails, it's foundational to the chaotic whole, like the jagged brushstrokes in a Riopelle painting. The sounds hold the city together. Without them, it would be a blank canvas.

But bats rely on an unspoiled acoustic space to paint their portraits of the world, one dot at a time. In the city, the constancy of noise pollution interferes with their capacity to do exactly that. Imagine that the pointillist painter Paul Signac was momentarily blinded by a flashlight while finishing his masterpiece *Antibes, le soir* (rolling hills, an anchored French schooner, the setting sun casting speckled notes, a chromatic scale of pastels). The painting would be rife with strange miscues: black amid a pink cloud, points of muddled red contaminating the schooner's otherwise green sail, the scene robbed of its coherence. This is what bats face in the city.

Biologists describe the sounds that bats make as "signal design," a sonic architecture that varies according to the specific function it's carrying out. While in flight, bats send out constant frequency calls, which are short and regular bursts of echolocation that make their auditory processing exceptionally precise. When they're hunting, the pattern switches: A precision maneuver like catching a fast-flying insect requires a rapid and uneven series of "frequency modulated calls," sent out in high-frequency bursts to aid the bat in pinpointing its prey. The trade-off is that because they oscillate faster, frequency modulated calls also burn out more quickly, so precision is limited to the area closest to the bat. The auditory range of bats varies between species, but generally stretches from 20 kilohertz (at the upper edge of the human auditory range) to 200 kilohertz, meaning the auditory range of bats is about ten times our own. Across this wide range, bats take advantage of the differences between low- and high-frequency waves to paint a composite picture of the world that can, at a moment's notice, zoom out to produce a panorama or microscopically enhance a single feature for clarity.

The nine hundred or so bat species that rely on echolocation all tailor their signal designs to the particulars of their homes. Signal designs that help the trident leaf-nosed bat adapt to the stark open landscape of the deserts of Tunisia are far different from those that help the silver-haired bat hunt with such ruthless efficiency through the complex mazes of North American forests. Then there are cities. And while there are more bat species living in urban environments than any other vertebrate synanthrope, only 83 of the roughly 1,500 known members of the Chiroptera order—6 percent—have found a foothold among us. Chiropterologists (common name: bat scientists) have termed that silent majority of bat species living in the wild as "urban avoiders": These are species whose echolocation patterns and flight styles are suited to dense forest because they echolocate at higher frequencies, providing them a more granular snapshot of nearby objects, which is very useful if you need to navigate through tree branches. (They also tend to have larger wings relative to the rest of their bodies—known as low wing loading—which improves their maneuverability around objects in tight spaces but makes them susceptible to wind gusts, which are far more rare under forest canopies than in an urban boundary layer.) At the other end of the spectrum are "urban exploiters": fast-flying bats that use lower-frequency echolocation, which allows their calls to travel the greater open distances of cities, and have slimmer, more aerodynamic wings that, like a jet plane, lack maneuverability but are less affected by turbulence, which is a constant in urban heat islands. But as biologically primed as urban exploiters might be to thrive in cities, they still have to contend with urban acoustic environments that are downright hostile.

Human intelligence has crafted instruments and machines that have been producing ultrasonic waves since the late nineteenth century. More than a hundred years later and the study of ultrasonic noise pollution is still in its infancy. In fact, it was only in 2016 that the first attempt was made to explore how deeply the various mechanical devices we've affixed in public places have immersed us in ultrasonic noise. This tracks, of course: If we can't sense something directly, there's little impetus to go out and describe it. Nevertheless, when scientists did finally start surveying, they were confronted with a whole new acoustic landscape rife with pollution. Ultrasonic noise pollution, it turns out, is very different from that produced within the infrasonic range:

Instead of sharp and irregular noises, it is a continuous stream of sound generated by a grab bag of technologies. Motion sensors. Public address systems (which are continuously self-monitoring to ensure they can receive and broadcast signals in the face of disasters). Pest repellants. Alarms. And—the most ubiquitous of all—smartphones. What's more, infrasonic sound sources often produce ultrasonic overtones, like the high-pitched resonances floating disconnected over the low growls in Inuit throat singing. Imagine a bad case of tinnitus blocking everything out, an unending ringing that wrestles the rest of the world into silence, and you have a sense of what we've done to the ultrasonic range. What's more, this constant wall of sound is most prevalent at night, which is exactly the time when bats emerge from their hibernacula and need unimpeded communication to hunt. When confronted with walls of human-produced ultrasonic noise, some bat species end up reducing the frequency of their calls by as much as 70 percent, the equivalent of hunting in near blindness.

It's not just ultrasonic noise that bothers bats, though. In the acoustic urban environment, traffic noise also causes bats to hunt less efficiently, despite the fact that it doesn't interfere directly with ultrasonic communication. This is because some bat species will switch from echolocation to a listening mode called gleaning, in which they rely on the sounds that ground insects and other prey emit—the long percussive chirps of crickets, the frogs' ribbits—to isolate the location of their next meal. The steady roar of traffic obscures those sounds, leaving bats with fewer weapons in their arsenal to catch prey.

This combination of infrasonic and ultrasonic noise pollution has forced urban bats to make a drastic adaptation: switching up their songs. Kuhl's pipistrelle, a palm-sized vesper ("evening") bat, is a highly social creature and one of the most successful synanthropes in the world. Abundant in Italy, Kuhl's pipistrelles live in colonies of about a hundred, which they leave to hunt insects on the wing. Once their chirrups and squeaks have isolated their next meal, the bats swoop in, using an acrobatic maneuver called aerial hawking, in which they somersault through the air and use their tail and wings as a pouch to catch their prey. To coordinate attacks and ward off competitors, the pipistrelles also emit low-frequency calls that cross into the infrasonic, meaning that if we tune our ears carefully (and we're lucky enough to find ourselves in Italy), we can hear them: short cries that drop rapidly in pitch. Remarkably, when Italian scientists compared pipistrelle calls in Naples (a bustling seaside metropolis of about a million people with an abundance of old buildings, in the shadow of Mount Vesuvius) and Benevento (a tranquil nearby town of about fifty thousand set on a hill and surrounded by rolling farmland), the urban calls from bats in Naples were significantly lower in frequency—at around 15 kilohertz, well within the range of human hearing. It's a curious adaptation to the hustle and bustle of the big city. Why would urban bats shift their calls downward to a frequency range already deeply polluted with human-created sounds? The reason, scientists think, is that the Neapolitan pipistrelles are using lower-frequency sounds to better map out the wide-open spaces of the city, as these travel farther than higher-frequency calls. What's more, because lower-frequency waves are more flexible, they're able to overcome the random assortment of architectural obstacles they might wind around in urban areas. In the case of Naples, those obstacles include a medieval castle, the headquarters of NATO's Allied Joint Force Command, the ruins of city walls dating back to the sixth century B.C.E., and a grid of narrow streets organized by Hellenic Greeks almost three thousand years ago. As the city rose beneath their wings over thousands of years, the pipistrelles have found ways to better navigate this anthropocentric environment, draping lower-frequency calls like long tapestry threads to create an image of the wide expanse of the urban acoustic space. It's artistry in motion. And it might be creating a whole new form of life.

■　■　■

After we visit her office and grab a bite to eat, Hwang invites me to stroll through her neighborhood. As we start walking, she pulls a small metal box out of her bag. It's an ultrasonic receiver, which passively collects high-frequency sound waves and downshifts them into a range within human hearing. Hwang turns it on, and it emits crackling white noise; as we walk, she fiddles with the frequency band, trying to land on a range that might catch echolocation calls coming from bats hunting overhead.

We chat about the best local pizza spots, the smell of the metal shop at her university, and urban animals. When I mention that there could be as many as 640,000 raccoons in Toronto, she looks stunned. "There are," she says slowly, "more raccoons in Toronto than there are people here in Buffalo!" Now it's my turn to be stunned. Hwang then mentions the #mprraccoon, a viral news story about a resident raccoon in St. Paul, Minnesota, that scaled a twenty-five-story UBS bank building over three days back in 2018. True to form, what drew Hwang in was the building. "The raccoon was only able to make it up," she explains, "because the tower was an older build that used textured concrete and had deep window ledges on each floor, so it could grip onto the imperfections in the walls and rest as it climbed." If it had been polished concrete or glass, she says, there would have been no story at all.

"It seems," I say as we wander along a poorly lit avenue, "that when animals are more visible, people have more of a problem with them. But your theory is 'If I make the objects that are housing these animals beautiful, that increases their acceptability, even though it also increases their profile.'"

Hwang nods. "Look at the kind of attention that gets paid to birdhouses. There are so many that are lovingly made; people spend hours painting them. And there's a certain level of care that you can sense in those artifacts," she says. "Projecting that same sense of care in my work is something I'm really interested in achieving." It's an achievement that's required her to evolve her thinking and diverge from traditional architecture into a new multispecies design space where the rules haven't yet been written.

Ten minutes later, we turn back up to Hwang's apartment building. The street is lined with large trees, but Hwang spies an opening in the leafy canopy up ahead. We walk toward it, and as the branches give way to a warm August night and a slate-colored starless sky, amid the sounds of traffic and human interaction, we hear it. Hwang and I both look down at the ultrasonic receiver with unfettered awe. It's a short knocking sound, rich and resonant, repeating every half second in a steady rhythm. Hwang raises the receiver and its tiny antennae to the sky, and I stare up at the light pollution, smiling stupidly. Suddenly the knocking speeds up precipitously and gets much louder, a sign the bat is shifting from constant frequency calls, which it uses for navigation, to frequency modulated sweeps, for hunting prey.

It's an odd thing to commune with an animal purely through sound. But then again, there's nothing strange about it for the bat. In the resonant knocks there is an obvious intelligence, and the receiver that Hwang places in my hand allows me to hear it circling above, sometimes close by, sometimes fading into silence. And while we cannot see it, the bat can see us: Its calls are bouncing off our human bodies and returning to their source, making us visible to our distant mammalian cousin flying in the darkness. In that moment, I find it deeply comforting to know that the sky isn't empty, even if it looks to be. The same is true for the tens of thousands of vacant lots across this city, which have been transformed from monuments of economic failure to ideal urban habitats for synanthropic bats that survive by speaking in new ways.

As you get older in the city, its mysteries tend to fade. Mind-blowing new restaurants close, and new mind-blowing ones soon take their place. The last-remaining underground techno club is shuttered, until the next last-remaining underground techno club opens.

Your exes, burning brightly at first, orbit your social circle in ever-widening ellipses until they are lost to another social universe altogether. On the first pass, it's thrilling and sometimes scary. After that, you're left with the realization that this has all happened before. But connection always feels new—be it with another person, to an animal staring at you with fixed eyes from across the evolutionary divide, or through the calls of invisible winged mammals flying up above.

We stand there awhile, both of us mesmerized by the fading in and out of the knocking as the bat circles back and forth above our heads, zeroing in on its prey. "When you hear a bat coming in live, it's just magical," says Hwang. "But it's also very unpredictable when they'll turn up." What it takes, I realize, is vibrating at a different frequency.

▪ ▪ ▪

The shifting of animal calls to fit specific environments is known as the acoustic adaptation hypothesis, and it's one of the key strategies that bats have used to exploit cities. Its evolutionary consequences, though, are only now coming into view. In Berlin, common pipistrelles (grape-sized bats with a range straddling the United Kingdom and Korea) have radically altered the way they communicate with one another. They are urban chatterboxes, making social calls at ten times the rate of their rural cousins. They also pitch their voices to a higher frequency than pipistrelles elsewhere and ratchet up the complexity of their missives. Scientists believe that the Berlin pipistrelle dialect—fast-talking, high-pitched, and complicated—is a way for the bats to negotiate with one another when they show up at the same large urban spaces where insects breed and congregate. The complex local dialect is also an effective way, scientists have found, for Berlin pipistrelles to intimidate competitor bat species and thereby claim food and territory for themselves. What is still unknown, though, is how these changes happened. The unique signal design of urban pipistrelles could simply be a behavioral adaptation to a new environment, passed down through generations. It could also be caused by selection pressures, whereby certain bats that happen to sing a little differently are naturally better adapted to finding a niche in Berlin. Whatever the cause, these changes in signal design portend the start of a monumental shift in bat biology.

Bats use social calls not only to share information but to share genes as well. Like birds, their mating behaviors are intimately connected with the signals they produce; if one bat sings a different song than another, they won't mate. The changes in signal design that Berlin pipistrelles have undertaken to adapt to the city are extreme—so extreme, in fact, that they might be acting as an acoustic barrier dividing the urban population and the rural bats that surround them from breeding with one another.

The emergence of a new species, known as speciation, is an evolutionary process that happens, in nature, over millennia. It's been exceedingly rare, then, for humans to observe the moment when a species branches off from its progenitors and becomes something the universe has never seen before. But just as cities warp the way that animals experience time, they also compress how much of it is needed for evolution to play itself out. Rather than hundreds of thousands of years, scientists now believe that species are emerging in urban centers within the span of decades or even sooner. One way this is happening is through "assortative mating," which is the tendency of animals to breed with those that resemble them, both physically and in the shared language they use to communicate. Among Mehely's horseshoe bats, females have been shown to be attracted to males who echolocate at higher frequencies, because higher-frequency signals are, as the scientists put it, "an honest signal of body size." The larger the bat, the higher its echolocation frequency; as a result, male bats of the species tend to sire more offspring than those that can't hit the high notes. But Berlin pipistrelles pitch their songs lower to survive in the city, so these downward-frequency shifts could be deterring rural females to mate with the Berliner males, effectively cleaving the two populations, rural and urban, apart forever.

Another way that urban bat species might be emerging is through differences in the timing of breeding cycles. Temperature changes trigger when bat species breed, and urban heat island effect in cities radically alters temperature. This has led to urban Egyptian fruit bats giving birth two and a half weeks before their rural counterparts. This might not sound like much of a difference, but their mating periods last only one month, usually from mid-December to mid-January. With such a short breeding period, a shift of two and a half weeks leaves scant overlap—just a week and a half—for the breeding cycles of urban and rural Egyptian fruit bats. If this trend continues, through accelerating climate change and greater temperature disparities between urban and rural areas, the mating periods of the fruit bats will stop overlapping completely, creating distinct populations traveling along their own evolutionary paths. It's a phenomenon being sung into being at full volume: the birth of a new species, brought to life by the city itself.

Chapter 3

The Soft Intelligence

Eurydice, dying now a second time, uttered no complaint against
her husband. What was there to complain of, but that she had been loved?

—Ovid, *Metamorphoses*

Synanthropes Thrive in Apocalypses. As we learned from synanthropic
bats, cities impose such profound influences on species that they may
even be sites of speciation—the creation of new forms of life. There is a
darker side to cities, though: They are deeply polluted, and sometimes
include microhabitats that aren't just dirty but corrosive, leaving animals
few opportunities for survival. Still, some unlikely synanthropes have
found a way to thrive amid the toxicity. So how do urban animals
find their way in habitats all but left for dead?

■ ■ ■

It's noon, but as black as night—darker, in fact, than any moonless
night I've ever encountered. I follow the thick rope, covered in brown
and green living things, pulling myself forward as if my life depends
on it, the void encroaching on all sides. There is no sound except my
breathing, which comes in gulps and ragged exhalations. My lizard

brain is wholly in charge, and for once I don't need my superego to explain what's happening: If I let go of the rope, I will be lost forever. I follow the shape in front of me, barely visible, a quiet mess of limbs, dressed in black too. *Follow the rope and follow the man, and you will survive.* Finally, I reach the end of the line, the rope tied unceremoniously to some kind of concrete slab, which has been left and forgotten like everything else here.

Motes of particulate swirl all around, enveloping me in an opaque cloud. I squint, but the detritus is so thick that I can barely make out shapes five feet in front of me. Now the panic, which I've kept at bay, is setting in. It's as if the darkness is anticipating every single movement I make, rearranging itself to keep me enveloped every time I try to break through. I grip the end of the rope, looping my fingers through the metal ring to which it is tightly knotted, and fish around in my vest for my flashlight. *Light.* But as soon as I turn it on, my heart sinks. The light is more blinding than the darkness. It reveals nothing, simply reflecting off the cloud that surrounds me, and making it even thicker. Then Ed Gulleksen, the shape in front of me, reaches with one hand toward a complicated bulky contraption attached to his other wrist like a tamed bird of prey. With a flick of his finger, a great floodlight illuminates the space, and for the first time since I've entered, I can finally see the terrain that we have submitted ourselves to.

Giant rusted I beams lie flat and discarded all around us, the remnants of some long-forgotten megaproject. Others stick up out of the silty ground at strange angles, their ends broken and splintered, the metal blistered and flaking away, the red and white of a cooked crab shell. It's a landscape of iron and mud. Piled unceremoniously amid the beams is some kind of hollowed-out structure, human-made, that I can barely make sense of, also covered in a thick layer of sludge. It's about the size of a van, with long wooden boards running along its sides, cracked and rickety. I reach out to steady myself against them, but they crumble to the touch.

Ed ushers me forward and we head southeast, following the

I beams and trash, staving off the void in our little bubble of light. We come across giant logs stacked stochastically like long grasses in a bird's nest, which I'm forced to pick my way over and under as I blindly follow Ed through the murk. I am utterly lost—if I were alone, there is no way I would get out of here alive—but Ed projects an unwavering calm. He has been here hundreds of times, he assured me before we entered, and knows the terrain by heart. I believe him. I have to.

Past the logs, we come upon the lip of a sheer drop. Ed runs the floodlight over the gap and the darkness parts; for a moment, I can see twenty feet below us, and below that, nothing but darkness. Without a moment's hesitation, Ed advances, scaling the cliff down. I take a breath and awkwardly follow along, picking my way carefully along the holds Ed used. The cliff is covered with multicolored plants and populated by small obscure animals that scurry away as my shadow falls over them.

A few minutes into the descent, Ed stops and signals me to come closer. His floodlight illuminates a network of foot-long breaks in the rock that open up into a complex cave structure. This must be the place. I follow the beam of light as it bounces off the cave's rough-hewn walls. Tumbling down the cliff face is a midden of discarded animal parts in a trail that leads straight back into the den. I peer inside and see bits of flesh and chitinous armor disappearing into the blackness. At that moment, the fear leaves me and only one thing becomes important.

I want to see the hunter.

▪   ▪   ▪

It was 2014 in Seattle, Washington, and Eliza Heery was at a loss. She had done the things normal people do: gone to school, tried to be kind, and moved through the world exacting the minimum level of damage possible. It hadn't been enough. The strangest part was how rock-solid it all had felt before the break.

In a world full of lonely people, Heery had somehow found a partner who had committed to a life with her. With a master's degree in hand, not to mention her partner's salary, the financial stress she had felt throughout her life had begun to ebb. And then, in what felt like one of her luckiest moments, Heery found an absolute steal of a house, a modest Seattle Box Modern–style structure priced far below its value, a physics-bending singularity in one of the most notoriously cutthroat real estate markets in North America. Sure, it needed new gutters, drainage, an overhaul of its foundation, and earthquake upgrades—but those were problems she felt lucky to have. Heery could exhale. Real happiness was right there in front of her.

And then came the shift—a moment of existential violence without warning—and Heery found herself completely unmoored. It began when her partner, with whom Heery had assumed she'd spend the rest of her life, left. He just called it quits, she says, with no real explanation. The loss of the relationship stung on an emotional level, but she didn't have time to fully process her grief before the greater implications of the break quickly set in. It was simple, really: Without this person, the life Heery had painstakingly built for herself was no longer financially possible.

Heery was enrolled in a PhD ecology program at the University of Washington that was heavy on fieldwork. For weeks on end, she was assigned to Friday Harbor, a quiet town of barely two thousand people on San Juan Island, amid an archipelago located off Washington's coast. These are mostly wild and rugged islands dominated by cedar and arbutus forests, over which ravens and bald eagles scream and soar. As a marine ecologist in training, Heery spent her days wading and diving in the pristine, salty waters of Puget Sound, one of the most biodiverse aquatic spaces in the world. It was beautiful, a dream job, but it came with a literal price: Without her partner's additional salary, Heery couldn't afford the mortgage on her home while also living on the island, a four-hour drive and an hour-long ferry ride due north of Seattle. Something had to give.

Years earlier, in high school, she had worked as an intern at the California Academy of Sciences in San Francisco, a sprawling complex in Golden Gate Park that hosts a planetarium, a terrarium, and an aquarium, the last of which is home to more than sixty thousand marine organisms. It was Heery's job, as the only intern that could scuba dive, to keep the aquarium's various tanks clean. That meant scrubbing algae off the plexiglass walls of the shark tank while the predators circled lazily right behind her, their unblinking eyes considering her at every pass, their currents rocking her gently as she worked. It also meant jumping into a massive donut-sized tank that housed hundreds of yellowtail tuna, which moved through the water at torrential speeds. A vitamin deficiency could cause the tuna to develop lockjaw and be unable to eat, so Heery was tasked with injecting them with vitamin E. It was complicated work, and not entirely safe: As soon as she descended into the water, she would be confronted by a dense, fast-moving school skittering all around her. As she tried to snare them, the tuna would accelerate to evade her and end up head-butting her in the face at high speed. It was an early lesson in how absurd and unpredictable intersections with wild animals could be, which is exactly why Heery was determined to embark on a career immersed in the waters of the Pacific Northwest, accruing knowledge and communing with the creatures that lived there.

And she came close. But what appeared in the moment to be a real estate opportunity too good to pass up became, after the abrupt departure of her boyfriend, financial handcuffs that kept her shackled to the land. The house demanded too much of her to leave for weeks on end; nor was there a way for her to supplement her income while living at a marine research site in the middle of nowhere. Marine ecology, it seemed, was for people with means: rich kids who could travel wherever they wanted, who lived in a rarefied world beyond her own. For Heery, it was just a fantasy. And it had ended.

■ ■ ■

"Somebody once told me," Heery says, "that when you're in the water with large fish, you have to look at them like you're looking at a cute boy. It's strange, but it's true: You can never look them straight in the eye or they freak out." She laughs, and so do I. "Which is so weird on so many levels." The ocean, Heery has discovered, is way weirder than she could have ever imagined, even dating back to her days dodging headbutts from lockjawed tuna. It's also, she's found, more generous and resilient than we have been led to believe, even in the places that humanity has left for dead.

Heery looks back at that moment of despair, when she realized that a fixed income could never get her to Friday Harbor and all of its splendors—kelp forests, orca pods, schools of squid, and absurdly vibrant and variegated soft coral colonies whose forms push the boundaries of the baroque—as an axial moment that shaped her life in profound and positive ways. Never someone to sulk, she sat in her new home, alone, and took stock of her options. She loved her dog. She loved her family, who had instilled in her a deep confidence in her capacity to survive. And she loved her West Seattle neighborhood, which boasted great coffee shops, larger-than-life industrial malls, and a few pebbly manufactured beaches where gray stones met gray water, which spilled off the busy gray concrete byways.

It was around then that living in Seattle began to confuse her. "It's a marine city," she says, "and there's water everywhere you look. And yet, so much of it is not the type of habitat that people really care about. And then there's the pollution." Heery explains that there are sixteen Superfund sites in and around Seattle, locations so contaminated with toxic waste that tens of millions of federal aid dollars and decades of time are needed to clean them up. According to the Washington State Department of Ecology, the waters surrounding the North Admiral neighborhood in West Seattle, where Heery lives, contain ammonia, arsenic, mercury, polychlorinated biphenyls (PCBs, a highly carcinogenic chemical long used as a coolant), and dozens of other dangerous substances. "Hundreds," clarifies Heery. "A lot of lead and mercury, but some other really complex chemicals embedded in the sediment. When you stir up the bottom, that's when you really run the risk of bathing yourself in all sorts of nasty things."

This isn't just a case of chemicals causing skin rashes or short-term illnesses for swimmers unlucky enough to be exposed to them (though that does indeed happen here). The pollution is so bad that nature's fundamental rules are being rewritten. "The organisms that it's really affecting are orcas," explains Heery. "Like, the firstborn of any female orca tends to die because of the level of PCBs in their system." When she says it, it's like a punch in the gut, and I let out an audible gasp. "I know," she says solemnly. "Yeah, and all mammals do this. You off-load half of your body's toxins every time you have a kid. I remember the first time I read that. I was, like, 'Wait—what does that mean for humans?' It's super overwhelming."

One day, to keep her sadness at bay, Heery ran. Starting at her home, she cruised in the direction of the water. To her right was Seattle's Industrial District West, the armpit of the neighborhood, a massive port ready to receive shipping containers painted in bright orange, red, and blue, surrounded by barbed-wire fencing, above which gantry cranes squatted high in the air. To her left was the Lockheed West Seattle Superfund site, forty acres of ordnance-related contamination that stretches across the land and the surrounding waters of Elliott Bay. Heery pushed her pace as she ran past the steel mill down the street from her house before crossing onto Harbor Drive, a narrow thoroughfare snaking along the water's edge. The whole industrial neighborhood suddenly opened up before her and she was confronted with a view that never got old: the unobstructed skyline of downtown Seattle glowing luminously across the bay. It was far away from the intertidal area at her feet, which was encrusted with piles of riprap and massive boulders purposefully placed there to wall off the waves, the salty sea lapping against an entirely artificial shoreline. Heery stood for a moment, taking in the scene: the light playing against the distant skyscrapers, the waves testing the ersatz beach, the waters revealing nothing.

"I remember that moment specifically," Heery says. "It was a really beautiful day in the early fall, before the rain had started, and it was so warm." The fall turnover was yet to happen, when the surface of the bay cools and sinks to the bottom, sending the nutrient-rich waters from the bottom up in a strange cascade that causes microscopic phytoplankton to bloom. When they do, it's as if a shroud has been cast across the swelling sea, the visibility receding to nearly zero. Despite being hyperaware of how toxic and polluted it was, Heery looked out at the water, awestruck. She laughs, remembering the moment. "Sometimes you just really want to get in." When I ask Heery to describe what it looks like, her voice softens. "My favorite is when there's this grayish teal that appears sometimes, especially where there's a lot of shell hash—it looks almost tropical."

But another revelation struck Heery about her surroundings that day. The entirety of the coastal area in her neighborhood was artificially constructed. To her right, the flat and ugly concrete of the port was sticking awkwardly out into the bay. The rocky beach in front of her was covered with boulders, stones, concrete, and sand imported from elsewhere to break the tide. To her left, a pier supported on rough-hewn, oil-painted logs cut across the shoreline beyond a parking lot; amateur fishermen stood at its edge, casting lines for salmon and lingcod. None of it was natural. And yet, Heery knew that the same ecological functions she had been studying up in Friday Harbor, in the pristine San Juan Islands, were present across the artificial structures of the rock wall, both above and below the ocean's surface. All of a sudden, as she watched rats scurry in and out of the rocks, she was filled with a kind of dismal hope. "*Maybe I can just do it here,*" she remembers thinking, "*even though everything else in my life feels like it's falling apart.*"

We think of cities as terrestrial places, rooted in the earth, growing up toward the sky like cactuses and across the land like weeds. But most of the world's largest cities are built on coastlines, and the urban population density of humans living on coasts is accelerating far faster than that of cities inland. Water is as central a feature of most cities across the globe as land is, and a city's first function is to control water and the pollution that we fill it with. In the rare instances when we think about the complex underground structures designed to fulfill this task, we generally picture them as a kind of void: featureless, unknowable, and impervious.

In Seattle, this myopia has been taken to extremes. Beyond the chemicals and radioactive elements leaking from Superfund sites, scientists have trawled the city's harbors with nets and discovered a world teeming with Styrofoam, insulation material, nurdles (lentil-sized pellets that are melted down to produce most plastic products), plastic bags, cigarette butts, broken glass, synthetic wires, and bits of old metal. In just a single square meter of a Seattle beach, there are an estimated 1,800 pieces of plastic just waiting to be drawn into the ocean and the open mouths of gulls, fish, crabs, mollusks, and the many other creatures of the deep. What happens below the surface, though, is always out of sight and out of mind.

Except Heery knew exactly what life was like in the abyss, having spent years diving in Seattle's waterways. She had seen what the city's forgotten trash became after it was unceremoniously heaped upon the ocean floor: a new kind of landscape, alien and toxic, that would, with enough time, become the death of a creature or its next home. As much as the debris was a problem, she realized, it was also part of an ecological continuum. And that meant the marine environment at the edge of her neighborhood, though completely human-made, wasn't devoid of hope.

Heery had picked up subtle signs of change. At the edge of the former Superfund site she passed along her run, there was a "crazy little park" by the water that had been restored as part of that site's cleanup. Jack Block Park was built on a spot contaminated by wood-treatment chemicals beginning in 1909, and where Lockheed Martin, the U.S. defense contractor worth hundreds of billions of dollars, had subsequently spent decades building weapons of war, spilling arsenic, lead, and mercury into Elliott Bay in the process. After that level of devastation, the bay was believed to be effectively dead.

But something magical was happening—something that Heery, undeterred by the swirling chemicals in the water, had seen with her own eyes during a dive. (Diving here requires being covered head to toe in a drysuit, a necessary protection against the cold water but also against the liquid mélange of heavy metals and other contaminants.) "There's this old dock," Heery remembers, "and right off of that dock a kelp forest was starting to form." Kelp forests are some of the most dynamic ecosystems on the planet, turning silty and barren ocean floors into highly biodiverse habitats that house all manner of species, including boxy and dour kelp crabs as well as nudibranchs, a family of psychedelically colored, shell-less mollusks. Heery also saw rockfish, sea urchins, otters, sea lions, and seals—species that all rely on kelp forests to live. But kelp itself is extremely sensitive to changes in water temperature and nutrient flows and can only get a foothold if conditions are perfectly aligned. After so many years of contamination, it seemed impossible that such a fragile ecosystem could have possibly emerged. But it was there, and it was growing.

"It's not the healthiest-looking kelp," Heery clarifies, "but the fact that it's even there is amazing." She pauses. "It's often tiny things that are hopeful. And in the broad scheme of things, Elliott Bay is the closest we have to a system that's hit rock bottom. And the fact that a kelp forest can persist there is incredible."

In the weeks after that run, Heery's mind was on fire. As she reflected on the tiny sparks of life that were emerging in the underwater garbage dump at the foot of her neighborhood, a door seemed to open for her. "I grew up in a city," she says. "Cities are what feel natural to me." Why, then, couldn't they also be nature itself?

Heery, searching for a doctoral thesis topic, came up with an idea. What if she studied the trash that had been thrown into the water around Seattle as if it were any other underwater structure capable of harboring life? Heery was aware of the research on synanthropes like rats, raccoons, bats, and coyotes, which tried to explain how these terrestrial animals adapted to their new homes. But when she started searching for studies on how underwater creatures adapted to trash, she found almost nothing. It was as if cities abruptly stopped where the water started. Heery knew better. During her dives, she had seen a familiar process repeat itself: manufactured objects became obsolete, were dumped into the ocean as trash, and eventually took on new roles populating underwater landscapes. The pollution in Seattle was extreme, and the litter covered practically every inch of the adjacent seafloor, so much so that in some ways the ocean had become even more urban than the land that bordered it, which at least had disguised itself with artificial green spaces to break up the gray drone of concrete.

On land, mesopredators—those in the middle of the food chain, like raccoons and skunks—are often best equipped for urban environments. That left Heery with a question: When the ocean is remade as a dark mirror of the city above, which mesopredator, if any, can adapt to the strange new world it finds itself in?

■　■　■

Ed and I hover over the clamshells spilling out of the cracks in the cliff face. We're six stories beneath the surface at Cove 2, at the edge of the former Lockheed Martin Superfund site in Elliott Bay. It's midsummer, and the visibility is atrocious—barely five feet—beyond which everything is just a murky black. Terrified of getting lost, I have been swimming so close to Ed throughout our dive that I keep having to dodge his fins, and not always successfully. The trail of shells is unmistakable evidence that a hunter lives here, a mesopredator, and one of the most intelligent species on Earth: the giant Pacific octopus, known by divers as the GPO. Both Ed and I slow down as we search the cracks for evidence that it is home. He shines his massive floodlight in while my dinky little flashlight does its best, probing for signs. A featureless pink blob, looking like a giant baby mouse, lies among the cracks below the midden. Like so much here, I have no idea whether it is a plant, an animal, or some strange effluent runoff. Whatever it may be, it is not what we are searching for.

My disappointment is palpable I'm sure, even in the darkness with a mask covering my face. Ed taps his palm with two fingers, the signal for me to check my air levels. I reach blindly for my gauge and pull it close to my mask: 1300 psi, less than half of a tank left. Wordlessly, Ed starts moving vertically and I follow him through the darkness, back up the cliff face, both of us surrounded by thick clouds of plankton.

We crest a rocky hill, a cairn-like mass that resembles a stack of bricks set in grout, Ed's floodlight flaring; I watch it recede as the darkness envelops us once again. Then, a couple feet in front of me, a cluster of enormous white sprouts suddenly appear out of nowhere, and I have to frantically swim backward so I don't hit it. It's a large log, it turns out, emerging out of the muddy seafloor at a diagonal, ringed on all sides by giant plumose anemones. These are sedentary creatures, animals disguised as plants, that sway like palm trees. There must be at least fifty of them, almost three feet high with long stalks and wispy florets of tentacles on top, all glowing a ghostly bluish white except for one cast in a beautiful deep pink, the color of grapefruit guts. We pass over the first log and another appears, also covered in a plumose colony, this one a mix of bluish white, grapefruit orange, and blood red. As I swim above them, my eye is drawn to one of the anemones, which has caught an egg-yolk jellyfish in its sticky venomous tentacles and is slowly consuming it, a process that can take a week.

We move toward the shallows and pass over the remains of a small powerboat, reclaimed by the ocean and crumbling to the touch. It's the *Honey Bear,* Ed tells me later, a wreck famous to local divers, which seeped battery acid into the waters here for decades (until a local amateur diver, fed up, removed the giant batteries herself in 2021). It looks like everything else down here: a resting place for silt, algae, sedentary animals, and salt.

I feel the tide gently rocking me, a sign that the surface is near. And as we follow the ascending shoreline, what was a barren landscape of silty rock and trash a second ago suddenly erupts into a kelp forest, living and fecund. Kelp crabs, squat and armored, sit idly on turrets above the forest like monks on barren mountaintops. Large brown kelp leaves surround them, interspersed with beet-red seaweed and the vibrant green of sea lettuce, illuminated now by gauzy beams of sunlight pushing through the murk. All sway together in unison with the tide, back and forth, long leafy strands gently tousled by the moon's gravity, the resistance of the water, and the subtle intentions of life.

And then, as shocking as a psychic break, we breach the surface and I'm frameshifted back into gray reality. I bob there for a moment and look across Elliott Bay at the downtown Seattle skyscape, the Space Needle extending up like a model rocket ship. As I make my way slowly to shore, heavy with diving weights and a now near-empty scuba tank, larger-than-life stacks of maroon, brown, and green shipping containers are being ferried across the port lands by the gantry cranes just a few hundred yards away.

On the beach waiting for us is Eliza Heery herself. She stands there, cherub-faced and smiling widely, with a pixie cut and a neutral openness about her, as if she could merge into whatever environment she finds herself in. If she wasn't eight months pregnant, she explains, she would have joined us underwater, but it's too late to dive without risking harm to the child, which will be her first. As she helps me offload my oxygen tank and weights, I can't help but remember her story of the stillborn orca calves, of mammals and firstborns and toxic loads, and I feel a hot shame pass over me.

■ ■ ■

When Eliza Heery tried to save her career by studying marine ecosystems that had hit rock bottom—one of which also happened to be at the end of her block—she had a vague notion that she'd find some interesting algae, seaweed, or small crustaceans living amid the trash that humans had dumped in Seattle's waterways. There was scant research on the subject. One study of the North Adriatic had shown that across four hundred kilometers of Mediterranean shoreline between Italy and Croatia, artificial underwater structures were more likely to house non-native sea squirts, minuscule invertebrates that spend their lives attached to rocks. Heery figured that her PhD research could follow that approach of focusing on tiny things, and make the case that their presence in urban marine areas was proof of how resilient ocean ecosystems can be to contamination.

When Heery first pulled on her scuba gear and walked into the chemically saturated waters that ring Seattle to start her research, she was immediately confronted with the true extent of the damage to the city's marine underbelly. The catalogue of trash was staggering. Amid brief patches of seaweed, small schools of fish, and clusters of marine invertebrates, there were rotting boat hulls, engine parts, and hitch trailers; buckets, barrels, oil drums, car parts, and an intact rusting bicycle with barnacle-encrusted tires; concrete slabs and rubble, cinder blocks, fiberglass containers and plastic, everywhere plastic; metal and concrete pipes, some as large as sewer tunnels; sheet metal and metal grating, piles of car tires and an entire van, fully corroded, folded into itself and barely recognizable; wooden and steel beams as long as ninety feet across; bricks, burlap sacks, flowerpots, and garden hoses. At a dive spot in her neighborhood, nicknamed the Alki Junkyard, there was even an intact garden gnome whose face had been entirely covered by a giant plumose anemone, the long-stalked carnivorous invertebrate glowing an iridescent bluish white. A little farther east, at the bottom of Cove 2, a plastic penguin statue stood sentry, not far from a pile of traffic cones, like an underwater satire of the bizarre world above. "The toilets," Heery confides to me, "are a little farther south." Elsewhere, Heery found handguns and cell phones, the hints of some interaction on the surface gone wrong. It was all so overwhelming. The ocean surrounding Seattle had become a place where the city's terrible secrets went to die. It was not where animals came to live.

But Heery's despair was short-lived. As soon as she began hunting the underwater garbage dumps around her neighborhood, she realized that something was off: The ecosystem wasn't acting as dead as it should have been. Instead, life—silent, secretive, tenacious life—was still there, haunting the turbid depths. The improbable kelp forest blooming in the underwater foothills of Jack Block Park was one early sign that marine habitats, even feeble ones, could recover and grow amid a toxic swirling mess. There were tiny hints everywhere, it turned out, that all was not lost. But there were also much, much bigger signs. "Diving up in the San Juan Islands," she says, where the clear, clean waters were an extreme counterpoint to Seattle's maritime mess, "you were lucky if you found one GPO." But in Seattle? "You can reliably find two to three per dive." Heery couldn't understand why—and confusion is the ideal place for a scientist to begin her journey. To her surprise and delight, Heery's neighborhood dives had prompted her to ask a question far bigger than she could have imagined: Could the giant Pacific octopus be an urban mesopredator?

Ecologists describe animal populations in many ways. One is by measuring their size; if animals in one place are physically larger than others elsewhere, it's a fair bet they're healthier. For more elusive creatures (including apex predators like tigers, pumas, and bears), ecologists rely on a much more personal approach: hunting, tagging, and tracking their movements. A third strategy is assessing their population density, which involves counting the number of creatures in a certain area (like the alarming rate of as many as one hundred raccoons per square kilometer in Toronto). Measuring an animal's population density is especially useful in understanding how different environments (like an underwater garbage dump) benefit specific species (like the octopus). To test her theory that GPOs were marine synanthropes, Heery was game for all three approaches, but she quickly realized the challenges of the first. Trying to measure the size of an animal that can whisk its way through the darkness in mere seconds, which can inflate and deflate its sac-like body depending on its mood, and which is constantly surrounded by a swirl of tentacles probing the darkness, would be ridiculously difficult. Previous efforts by marine ecologists to try the second approach—tagging octopus—also generally ended in failure. "They rip the tag out in minutes," Heery explains, regardless of how you fasten them on. "They're Houdinis."

"Cephalopods," she says with a laugh, "are hard." Among this class of mollusks, which contains not only octopus but also squid, cuttlefish, and the ancient nautilus (an order of living fossils that haven't changed in hundreds of millions of years), elusiveness is a constant, making them poor subjects of study. "It's especially hard with GPOs," Heery says. "They're so big, but they move so much, and while they have one main den, they'll have little lunch spots where they might hang out at other times." (I think of lingering in front of the octopus midden alongside Ed Gulleksen. Clearly, we had stumbled upon one of many lunch spots.) Further complicating matters is the unfair competitive advantage the giant Pacific octopus has in eluding the efforts of other organisms (humans, but also seals, sea lions, dolphins, sharks, and whales) in trapping it. With eight powerful arms that can reach up to twenty feet across and a decentralized intelligence that includes nine brains—one that controls each arm's movements and one inside its soft domed head that manages its central nervous system—the GPO has a multitude of evasion strategies at its disposal. Its eyes, and their rectangular pupils, are marvels of engineering, remaining horizontal regardless of the three-dimensional movements the animal makes. Like all cephalopods, it has only one type of light receptor (compared to the two humans have), which restricts their world to grayscale tones. But through a process called chromatic aberration, muscles surrounding each eyeball can squeeze it into different shapes, which in turn filters light into different wavelengths as it passes through its lens, allowing it to interpret color indirectly. Its permeable skin, embedded with pigment-filled sacs called chromatophores, can be manipulated to change both its color and its texture, while proteins embedded in its skin called opsins have the ability to sense light waves directly, a power the GPO uses to instantly blend into its environment—or to make its skin glow bright red if you get it angry. Each of its two thousand suction cups can grip ten pounds of weight, and those near its mouth can grip as much as thirty-five pounds each. (If you're ever lucky enough to be handled by one, Heery tells me, there is no scenario in which you can outmuscle it. Instead, you must wait it out, gently nudging its arms off your body, and pray that it doesn't try to pull your mask off or the regulator out of your mouth while you hang suspended sixty feet below the surface.)

When estimating their size and tracking them across their ranges proved impossible, Heery tried the last approach: measuring their population density. And to get a sense of where she should look, she turned to an army of citizen scientists who had combed every inch of Seattle's seafloor and were more than happy to kibitz about it. "When we first started the project," she says, "we went to every diver we could find to get their ideas about where to look. And always, it was like, 'Oh, there's this wreck here'; you know, 'Here's where they dumped all the debris from the Tacoma Narrows Bridge when it fell down in 1940.'" With tips on dozens of promising sites, Heery set about counting the number of giant Pacific octopuses living in her backyard. Soon, what had seemed improbable shifted into the surreal.

Octopuses are mesopredators, roughly in the same place along the marine food chain as opossums. This welterweight class is known to thrive in ecosystems where prey is abundant and predators few. In cities, that's part of a phenomenon called the "luxury effect," proposed by urban ecologists, which purports that there is a linear relationship between a neighborhood's median income and its level of biodiversity. The wealthier the neighborhood, the greater the abundance of animals; poor neighborhoods, in contrast, harbor only a few select species. As a result, urban mesopredators—coyotes and raccoons, opossums and bats, among others—go where the wealth is: rich neighborhoods flush with parks, large backyards, and ornamental gardens. Once there, they can thrive within these simulated wild habitats surrounded by a cornucopia of mice, insects, fruits, and other small and nutritious bodies that sustain them in relative peace. Heery had no reason to think that the luxury effect wouldn't hold underwater. If anything, given how valuable properties were in oceanfront neighborhoods that weren't dumping grounds for toxic waste, it was a good bet that a seaside luxury effect might even have more extreme outliers. Sea life would surely be abundant off the coast of unspoiled beaches and the multimillion-dollar homes that lined them, while shitty little oil-stained harbors like the ones near Heery's place would be as lifeless as the plastic bottle caps, cigarette butts, and imported pebbles that littered their edges.

But Heery discovered that the luxury effect wasn't translating under the waves in Seattle. If anything, the phenomenon appeared to be playing out in reverse. She was seeing lots of octopuses in underwater dumping grounds filled with rotting vans, piles of decomposing tires, and the occasional handgun. She found an octopus that had made its home in an overturned bathtub, and another one living in a 1970s-era refrigerator. Elsewhere, she discovered a wrecked boat that a playful octopus had filled with dozens of golf balls it had collected from across the seafloor. ("Why?" she asks rhetorically. "I have no idea.") To her eye, it seemed there were more GPOs living in heavily polluted areas than in more pristine parts of the ocean. But could it be that the octopus really prefers garbage to its natural habitats?

To test this hypothesis, Heery mapped dive sites across Seattle, classifying them based on whether they contained vast amounts of garbage or only very little. To do so, she carried out underwater video transect surveys, a technique that employs special video cameras fitted with two lasers, which run in straight lines across a landscape to test its topography. Heery then compared the number of octopuses in sites that were closer to Seattle and had larger junk piles with those that were farther away and weren't covered in trash. Finally, she built statistical models testing the relationship between these variables to see if she could predict where octopuses were most likely to live.

When she ran the numbers, the results were not at all what she'd expected. All the divers she talked to had told her that GPOs loved the areas close to Seattle's shoreline, and she had seen them herself living happily in garbage. But her model disagreed: The closer a site was to Seattle, the less likely it was to house octopuses. Worse still, the statistical models fit poorly, meaning the variables weren't a good explanation for why octopuses were choosing some sites over others.

Something was off. Could it be that she had been swept up in a local legend? When she thought about it, Heery had to admit that there was one major reason why it was all quite unlikely. The places she was investigating—intertidal areas close to the city—overlapped with the areas where other marine synanthropes that preyed on octopuses tended to hang out. As a mesopredator, the GPO has to worry as much about what it's eating as what might eat it. And across the Salish Sea, studies have shown that the bellies of pinnipeds (a subset of Carnivora that includes seals, sea lions, and walruses) are filled with GPOs. Pinnipeds are fond of urban coastlines, and unlike land-based synanthropic mesopredators, they often have no problem basking out in the open, day and night, outdoing the aggression of even the most fearless gaze of raccoons.* Urban pinnipeds are one step up the food chain from GPOs (though still below orcas, who hunt them voraciously). In response,

---

* Once they claim their space, urban pinnipeds are also excruciatingly difficult to dislodge. For a taste of what these barking, fin-slapping, fur-covered carnivores can do, you should head due south from Seattle to San Diego, at the United States' other Pacific border, a city with some of the most beautiful (and crowded) beaches in North America. Once there, you can watch as iPhone-toting humans pick their ways through colonies of sea lions to live stream them up close. In recent years, the feedback loop of more humans gathering to watch more of the half-ton synanthropes, who use the beaches as a breeding ground in the summer, has led them to attack human beachgoers, setting off mass panics and extended beach closures.

octopuses have tended to live and hunt farther out at sea, in secluded areas that attract less attention from the hungry mammals. Or at least that's what marine ecologists thought—and what Heery's model seemed to confirm, even if she had seen differently with her own eyes.

The poor fit of Heery's statistical model, though, was like a crack in a dam, signaling a deeper problem in her data that undermined its structural integrity. To improve fit, she had to revise her model, either by taking out variables that weren't explaining anything or by adding in a factor that might be importantly related to GPO behavior but was currently missing. As she reviewed the data, Heery considered that if GPOs were indeed moving into cities, they would surely have to adapt to their new locales just like a land-based synanthrope. And that meant finding a way to maximize their food sources while also reducing their exposure to predators. Unlike on land, urban marine ecosystems afforded their residents an extra dimension of movement: up and down.

In Seattle, harbor seals and sea lions own the shoreline, where they snack on fish, clams, and octopuses. "In the shallows," Heery says, "even humans get pushed around by sea lions and seals, but in my experience, they don't mess with us at more than forty feet or so." If GPOs were going to survive in such a hostile environment, she surmised, they would have to adapt themselves to living in deeper waters, where seals and sea lions rarely ventured. So Heery added one more variable into her model: an interaction term that measured the combined impact of a depth of seventy-five feet or more with a site's urban intensity, which she defined by looking at the surrounding area's human population density, the number of artificial barriers that had been built on surrounding shores, how many roads there were, and the scale of land development. The idea was to see whether octopuses, once they moved closer to Seattle, might change the depth of their dens to avoid urban predators, and thereby more successfully adapt to their new environments.

When she ran the model with the depth factor added, its fit improved immensely, a sign that the collection of variables she had chosen were better explaining why octopuses were living near Seattle's urban areas. What's more, the interaction term—depth and urbanization—was highly statistically significant and flipped her previous results on their head. Octopuses, she found, much preferred living in urban sites compared to more pristine ones, but only at depths of seventy-five feet or more. Heery had unlocked the mystery.

Seattle's GPOs, which in the wild stuck to reefs and the rocky underwater cliff faces found in shallower waters, had learned to dive deeper into the murk as they entered cities. At those depths, the ocean floor in the pristine areas of Puget Sound is flat and silty, an endless horizon without hiding places. As soon as human debris hits the water, though, it transforms into something else: a landscape feature, like any cliff face or rock pile, that marine creatures instinctively populate. Trash in the depths, where shelter is so rare, opens up an entirely new oceanic domain for the octopus to exploit. For Heery, her findings brought her to an odd realization: All of the trash that had been thrown into the ocean, which had been slowly destroying it, was also providing this elegant and enigmatic animal with a new path to survival. A busted van is a nuisance, and throwing it into the ocean is illegal. But when it hits the bottom of the ocean eighty feet below the surface, it becomes a perfect hiding spot for an octopus, which otherwise would have nowhere to run when a predator came upon them, or have any way to surprise their prey. This is particularly important for GPOs, given that so many of their most impressive behaviors, like shifting their shape, color, and texture, rely on the presence of some topographical feature to match their camouflage against. Without objects to blend in against, they lose their competitive advantage. Urban marine habitats, Heery realized, provide an abundance of anthropogenic debris for them to use as dens, plenty of clams and other mollusks to eat, and an existence blessedly free of the pinnipeds that roam the intertidal shallows above. It was a once-in-a-lifetime scientific triumph: Heery had discovered that urbanization extended into the ocean, where it allowed one of the most intelligent species on Earth to thrive as a synanthrope.

I asked Heery what life is like for GPOs in these deep urban waters, living their life among forgotten trash. "I have seen them in tire wreaths," she says, "but they're not as common there. Concrete really is the top thing, and even if there's just one slab, they can dig out the underside of it and live under that." She pauses, mentally running through the various items she's seen underwater. "And old refrigerators, of course."

"That just seems totally dystopian," I say.

Heery shrugs. "I mean, it's a good life if you've got all the crab you need," she says. "And who knows—it's probably all really toxic crab, but GPOs only live for three years anyway, so, what's the worst that can happen?"

Heery telegraphs joy when she talks about the ocean along with a deep-in-her-bones love of the organisms and ecosystems that it contains. And yet, her words are filled with a cynicism bordering on the nihilistic.

"What is it about you," I say, "that you see a super-polluted ugly waterway and that attracts you?"

Heery considers the question. "Look, the water in the middle of the city is not really beautiful in the way that makes people think, *I'm going to be a marine ecologist*," she concedes. "The other big part of it is that I'm at a point in my career where . . ." Heery trails off, then takes a breath. "I go to conferences and hear keynote speakers, people who I respect so much, who have long warned everyone about how changes that we're making are going to have these really awful, vast consequences, and it's really easy for them, at their career stage, to move into the doom-and-gloom realm where that's all there is left to talk about. But after a while, just hearing that it's all really bad and all the changes we're making are terrible—which I don't disagree with—it isn't enough to motivate me."

"So grief is a luxury?" I ask. "This idea of having seen paradise and having seen it be lost—that's a luxury?"

"I think grief is really appropriate," Heery replies. "But the question is, what do you do with that? If you're coming into ecology now, you have to deal with the idea that the climate we thought we could have is not what it's going to look like in the future. And this"—she gestures to the polluted urban coves that surround her—"is the best place to look if we want to know what it will look like in the future. If there is any place to find hope, I think it's here. Because things still find a way to grow here."

▪ ▪ ▪

Two months after I first scoured Seattle's trash-covered seafloor for a giant Pacific octopus, I'm back in the water with Ed Gulleksen, this time in Poverty Bay, a modest body of water near Seattle's southern suburban edge. We're about fifty feet down and one hundred feet from the shore, where two nice ladies standing mutely at a Jehovah's Witness booth watched with disinterest as we made our way into the ocean, slowly paddled out alongside the short pier, and performed the miracle of breathing underwater. It's like night and day from our last dive: Instead of being enveloped in a tight bubble of darkness, the ocean is bright and transparent today, an open three-dimensional space inviting me in. Textured sunlight drips down from the surface, shifting the colors diffracting through the water surrounding me—green, yellow, orange, red—and imbuing it with a satisfying viscosity, as if I were swimming through the dissolved insides of a giant gummy bear.

After the drop, we settle on the seafloor amid intersecting cables that extend at acute angles far across the flat, silty landscape and out of sight. Ed immediately goes to work, and I follow as he leads me along a path above one of the cables. I frantically frog-kick behind him, my breathing ragged in my ears, big bubbles flowing madly out of my regulator as I try to keep up. We descend farther, to about eighty feet, the water cold and bracing, brain freeze setting in despite my neoprene hood. The ground here is as featureless as the rest of the seafloor, but it's at this depth that Heery's research shows that urban GPOs are most likely to reside. Ed assured me that diving in this Seattle suburb would be well worth my time. So far, though, all I have seen is flat yellow-brown mud punctuated with trash—a disappointing mix of the dead and the inert with which I am by this time pretty familiar. Unaware of my skepticism (which is, it turns out, hard to telegraph while wearing a drysuit and breathing apparatus), Ed continues to snake along the crisscross of cabling that has been strung along the bottom, the only landmark across this unending muddy expanse.

A few minutes later, the terrain shifts. Coming into view in front of us is a small assemblage, about four feet high. There's a rusted iron box sitting on a couple cinder blocks, the stack covered by two big slabs of concrete that look like broken pieces of sidewalk, which have been set to lean against each other like an underwater A-frame. The edifice is undeniably trash but also evidently took some planning and effort. The concrete slabs must weigh at least a hundred pounds each, and dragging them across the seafloor and into place to create the facsimile of a farmhouse roof could not have been the work of chance or just one diver. Surrounding the structure are a few other discarded cinder blocks and rusted pipes, over which a half dozen small brown fish, spiky and sequined, swim in lazy figure eights.

Ed makes a beeline for the structure, his floodlight flaring to reveal a gaggle of rock cod, thick-lipped and grimacing, moving quickly in and out of the otherwise empty iron box. From a distance, my heart sinks. Based on Heery's discovery, the edifice is tailor-made for an urban GPO: It's "anthropological debris," as she puts it, transformed into a landscape feature fit for the elusive mesopredator. If I am going to finally lay my eyes on one, after months of searching, this will be the place. Evidently, though, Seattle's octopuses are like city dwellers the world over, too busy cogitating in their own problems to care about my needs.

I hang back and watch Ed stick his face into the iron box, a steady stream of bubbles enveloping his head and rising to the surface. He's snapping photos of the rock cod for his collection, and I try to muster some appreciation for all the time he has spent dragging me around Seattle's seafloor. Still, I'm not motivated to join him at the box. I have seen lots of fish already, and unlike Ed, who gets excited about seeing a stubby and unremarkable flatfish lying on its left side instead of its right, the minute pleasures of the city's underwater junkyards are beyond my unrefined palate. I am ready to leave.

Ed pops his head out from the box and makes a wide gesture to call me over. Compliant as always (when you're underwater, eighty feet deep, and completely lost, it's best to obey the person who brought you there), I swim slowly over to him to examine the maroon and white-spotted rock cod up close. As I get closer, he pulls away, and then too do the rock cod, giving me an unobstructed view into the box. There, amid a mass of purple tentacles and perfectly circular white suckers, a single unblinking eye observes my approach.

Suddenly everything becomes very still. The bubbles have ceased to flow out of my regulator, and my lungs feel full to bursting. I realize I've stopped breathing. Meanwhile, the octopus's eye remains fixed on me as I settle down in front of it, my hands clutching the silty, goo-like seafloor in front of its purpose-built home. With an effort, I inhale and then let out a loud, hoarse exhale, a riot of bubbles obscuring my vision for a moment.

The iron box is about a foot and a half high and wide, and just as deep. The GPO crouches on its side, its tentacles held aloft against the rusted roof, steadying its body in place, its purplish-red skin vibrant against the white of its exposed suckers, which run in two neat rows in steadily increasing size as they approach its mantle. Its visible eye, about the size of a golf ball, is open, but barely; still, I can see its rectangular pupil, providing it with panoramic vision. A few inches lower down its mantle, its breathing is visible, as ragged as my own, flowing through a ruddy maroon sheath that opens to expose a perfectly smooth and eggshell-white orifice about the size and shape of a navel orange. These are its gills, which labor strangely to draw in and expel water through a short tube-like siphon hidden partially under its body. I feel the tug of a gentle current pulling at me, but I stay fixed on the seafloor, holding fast like an anemone. The GPO is barely moving and radiates exhaustion, each intake of oxygen seeming like it could be its last. I have seen that kind of breathing in people about to die, and it's strange to clearly recognize the end of life in a creature whose lineage separated from our own 750 million years ago, when amoebas, proto-sponges, and our shared ancestor, small flatworms, were the apex of complex life.

But there's more to its curious immobility than an impending death. Amid the tentacles it presses aloft against the walls of its anthropogenic home, under the cover of the two great concrete slabs, the giant Pacific octopus holds a clutch of eggs, translucent and ectoplasm green, which cascade down before it like long bunches of grapes, each about the size and shape of a raindrop. I run the beam of my flashlight along the clutch, careful to avoid shining it into the mother's half-lidded eyeball. In each of the individual egg sacs, I can see black spots—the beginning of the hatchlings' own eyeballs, which will guide them as they navigate through the waste-filled world that will become their home in just a few short months.

GPOs produce up to four hundred thousand eggs in a clutch, among which only one or two survive to adulthood. The octopus is among nature's most devoted mothers, tending to its one and only clutch over a seven-month development period, during which it stops eating; when all the eggs have hatched, it lies down and dies. The mother in front of me, though, is not as far gone as she seems. Aside from her hard breathing, the tip of one of her tentacles emerges from behind her body, and she begins to run it delicately along the eggs, up and down, the tender but firm movement of any concerned parent attending to its brood. Although they are among the longest-lived octopus species, the giant Pacific octopus still only has a lifespan of two to three years. For them, birth heralds the onset of death. The scene in front of me is like a sped-up version of life itself, creation and annihilation happening so closely together that they might as well overlap completely, the countdown to both coming with each breath, this pantomime of existence framed by human waste—the killer of the orca's firstborn—transformed into a home.

While I watch an expectant mother come to the end of her life, somewhere up above the surface Eliza Heery lies in bed with her own newborn. Both are acts of hope in the midst of an environment, the twenty-first-century city, that is the closest thing ecologists have found to rock bottom.

Chapter 4

Grave Architecture

They adapt quickly, breed quicker and have no predator enemies . . . except man. We need to stop abdicating our position at the top of the food chain. They need to be shot.

—Nextdoor post, Los Angeles

Cities Both Trap and Protect Animals. The urban giant Pacific octopus exemplifies just how unforeseen the effects of cities can be on ecological balance, as well as the adaptive power of synanthropes in overcoming habitat destruction. This phenomenon of mass ecological destruction isn't only playing out under the waves. Cities expand rapidly across land; they absorb entire ecosystems into their unrelenting sprawl, forcing animals—even seemingly ill-suited ones—to adapt to their new urban enclosures. At other times, animals are lured into cities by the promise of abundance, which might be real or might be a trap. These multiple pathways have led to an increasing variety of animals finding ecological niches in cities. Once ensconced, though, even the most highly adapted synanthropes face the possibility that the relative benefits of urban living might cost their species its survival in the long run.

■ ■ ■

Raccoons have elastic brains and prehensile fingers. Bats take to the skies and can see in the darkness. Octopuses are cunning and amorphous hunters that turn our trash into protective habitats. But there is another animal, awkward and seemingly ill-fitted to life among humans, that has somehow found an urban niche across North America: the coyote.

Coyotes are an odd addition to the pantheon of successful synanthropes. A member of the Canidae family of Carnivora and closely related to wolves, they don't seem to possess a clear separating feature (like a large brain, prehensile hands, or the ability to fly) that would allow them to thrive in the tangled set of microhabitats we call cities. They're also bigger than many other urban mesopredators (about the size of a German shepherd), which makes them conspicuous, and they're often loners, meaning their social learning is limited. They can't climb trees, and their diet consists largely of meat. They're also surprisingly skittish. And yet, among the fifteen or so species of Carnivora that have adapted to human environments, only four have mastered the entire range of anthropogenic habitats: farmlands, villages, suburbia, and city centers. Three of these carnivores are the raccoon, the red fox, and the stone marten, small and skillful animals that all weigh a little more than ten pounds, on average. The other is the coyote. Weighing up to forty-four pounds, with a long bushy tail, yellow eyes, and a long slender muzzle, it is the coyote's flexible relationship with power that has helped it thrive among human beings.

Over the last few decades, the world of the coyote has blown wide open and the animals have now colonized every major city in North America. In Chicago, long a haven for the canid, the number of coyotes removed from the metropolitan area increased fifteen-fold during the 1990s. Try to dig a little deeper into the history of urban coyotes, though, and you encounter a massive void. Ecologists have been studying urban animals for over a century, given the ubiquity of rats, bats, and the sweet songbirds that have long been drawn into urban heat islands. But there is surprisingly little in the scientific record about urban coyotes.

One of the earliest scientific studies of urban coyotes is a two-page write-up in the April 1980 edition of *The American Midland Naturalist*. It's a poignant story, written with the understated flair of one of Chekhov's peasant tales, tracing the movements of a yearling coyote over the year 1976 in Lincoln, Nebraska, after the animal found itself living in a new suburban development that had rapidly sprung up amid farmland and forest. The biologists, who collared and tracked the young animal, describe its territory as "grain sorghum fields in southern and eastern sections, and riparian strips of deciduous trees, shrubs, forbs and brome grass along the southern edge; two railroad rights-of-way dominated by forbs, grasses and woody vegetation intersected the home range." But it's the descriptions of a lone coyote simultaneously running from humans while looking to ease its loneliness that hits the hardest. Between January and May 1976, and then July through November, the "Lincoln coyote" (my name for this fascinating individual) was frequently observed playing with a female golden retriever in a suburban park. They would run and play-fight together, their sessions initiated by the coyote, who would gaze at the golden retriever and raise a forepaw in the air, aptly described by the biologists as a "meta-communicative signal." But their love was doomed: When the golden retriever was in heat, the coyote tried to mount her, only to be chased away by the dog's owner. Unbowed, the coyote tried again over the ensuing weeks, but a male retriever with designs on the female blocked it at every turn, forcing the yearling to retreat. At other times, bands of humans chased it into a culvert, where it hunkered down to wait until the threat subsided. But the coyote never retaliated with attacks on humans or their property. Instead, it contented itself to hunting and eating cottontails (a cuddly pest that had by then overrun wide swaths of Nebraska), as it tried and failed to make sense of the new urban labyrinth that had enveloped its natural habitat.

The Lincoln coyote is part of the early wave of synanthropic canids that experienced a fast-forward reordering of their homes as cities expanded rapidly. Ecologists sometimes describe the city as a matrix, simultaneously expanding in size while continually dividing itself into smaller fragments. In the 1990s, cities in North America expanded at an unprecedented rate, growing by almost 10 percent per year on average, though cities in the South and Northwest grew by an astounding 19 percent. That led to coyotes finding themselves abruptly enclosed within new urban landscapes from which they could not muster an escape, as relocating into unknown territory elsewhere presented too many uncertainties for this famously neophobic animal. In some cases, enclosure isn't just a metaphor: What once had been open grassland and forest was, over a period of weeks or even a few days, leveled and cordoned off, leaving an erstwhile, free-moving animal bumping up against impassable boundaries like fences, highways, concrete walls, and the identical houses of planned communities.

The Lincoln coyote was an unlucky descendant of the first *Canis latrans* (Latin: "barking dog"), our modern coyote, which emerged on the North American plains 1 million years ago. Back then, coyotes were much larger creatures conditioned to take down big game like deer while fending off more massive competitors like the dire wolf, which had a bite stronger than any other *Canis* species that has ever lived. About ten thousand years ago, when the dire wolf went extinct after the dying out of its megafauna prey like mammoths, giant ground sloths, and North American camels—a product of climate change, human competition, or both—gray wolves rushed to fill the void at the top of the food chain. In response to the ascendance of the smaller pack-based apex predator, coyotes evolved to also become smaller as well as more opportunistic, just as willing to scavenge on carrion as they were to hunt small rodents, rabbits, and insects. What's more, the coyote became a socially ambidextrous creature, content to hunt alone or join a small pack, moving fluidly between a life of isolation and one of social reward.

When the Lincoln coyote lay in the tall grasses at the edge of farmers' fields, when it strode through newly created backyards and residential streets, observed by horrified homeowners, and when it lay under shrubs and on carefully mowed lawns, it was the product of ten thousand years of evolution that had conditioned it to survive at the margins of apex predators. No matter whether it be wolf or human being, the lesson was the same: There is a life to be lived—a rich one, even—in the shadow of kings.

■ ■ ■

The Lincoln coyote, a loner throughout its short life, was nevertheless part of a larger movement. Before 1700, the home range of *Canis latrans* was restricted to the central band of the North American continent, from the Canadian prairies all the way down to the deserts of southern Mexico, bound by Appalachia to the east and the Rocky Mountains to the west. This was a time when wolves and bears still claimed sovereignty over the land, before a sustained campaign of mass extirpation left their populations decimated. Coyotes, true to their nature, mostly evaded the violence of humans but followed closely at their heels, moving into habitats suddenly devoid of apex predators, where they were free to rule. From 1700 on, coyotes expanded their range in all directions by about 55 miles per decade (though at some points, like in the 1950s, their range expanded by 125 miles), leaving the animals firmly ensconced across Alaska, from California all the way to Baja California Sur, Mexico, and even as far east as Cape Breton Island off the coast of Nova Scotia. Along with the removal of competitor species, some biologists have offered alternative explanations for their sudden expansion, including catch-and-release programs that left unlucky individuals in unfamiliar landscapes where they were forced to adapt or die, or the possibility that coyotes were always present in their current range but at numbers so low that their presence was overlooked.

There's a third alternative. Between 1630 and 1930, 361 million acres of land in the United States—15 percent of its forests and 20 percent of its grassland and scrub—were cleared and transformed into farms. By the 1950s, though, this cropland was being abandoned across the continent as the country shifted away from farming and toward manufacturing. This left vast swaths of abandoned terrain that had been remade to closely mimic plains and deserts, which are the coyote's natural habitats. It's almost as if our species has been working to systematically remake the North American continent into a coyote paradise, free of pesky competitors like wolves and bears, while artificially expanding its preferred habitat as far as possible.

▪ ▪ ▪

"Did you hear about the coyote in the Quiznos?" Chris Schell had prepared for a lot of tough questions during his interview for the doctoral program at the University of Chicago's elite Committee on Evolutionary Biology, but not this one. The interdisciplinary program was ranked among the best in the world for ecology, and Schell had figured the admissions panel would be made up of a bunch of smart ecologists mostly interested in preserving the status quo. So he came in with a plan. "*I'll propose something about wolves and African wild dogs,*" Schell remembers thinking, "essentially what people tell you that you have to do in ecology." A traditional thesis project would be boring, but he imagined that hewing to convention was the price of admission; the important thing was to get into the school. The truth—which he planned to keep hidden from the gatekeepers—was that his real passion lay much closer to home.

Schell grew up in a predominantly Black neighborhood in Altadena, a suburb of Los Angeles situated between the wealthier city of Pasadena and the rugged grandeur of the San Gabriel Mountains. When, as a teenager, he first told his family that he wanted to be an ecologist, the reply was blunt: "I have no idea what that means." He laughs. "Most folks in my community, the Black community in Los Angeles, when you say you're interested in sciences, they're, like,

'Okay, you'll be a great doctor.' Well, I tried that," he says, "and it sucked. Did not like it. Too sterile, not enough dynamism—"

"Not enough biting?" I offer.

Schell smiles. "Not enough biting!"

Though he grew up in one of the largest metropolitan areas in the world, his community also backed onto the Angeles National Forest, and it was there that he spent the most formative years of his life. The forest is 700,000 acres of protected land that includes vast swaths of Southern California black walnut, a sprawling, multi-trunked shrub-like tree that glows a brilliant red in the fall, along with old-growth white fir and ponderosa pine. Amid the scrubby mountainous terrain, under the protection of the leafy forest canopy, bear, bobcat, bighorn sheep, rattlesnake, and coyote move fluidly, occasionally overlapping in their orbits, the sight and smell of the other presaging a lengthening of their life or its abrupt ending.

The same forces at work in the forest behind his home—the tension between shared and segregated habitats—infused Schell's understanding of how cities function. "I grew up in Altadena before it was gentrified," he explains, "when it was majority Black. Fifteen minutes south of us in South Pasadena were these huge homes, owned mostly by white folks that had a whole lot of money. I intimately knew as a kid how different the ecologies were." Those urban ecologies ended up blurring with that of the forest, where Schell would spend his days wandering along the banks of placid streams and testing which ones he could jump across. After long afternoons spent immersed in nature, he would come home, only to find the roles reversed: Safely ensconced inside his family's house, he would look out to see forest wildlife, mostly coyotes, wandering through his backyard, as curious and cautious in his habitat as he had been in theirs just a few hours prior. Still, that feeling of permeability between the two worlds—a boy moving amid the coyotes, the coyotes then aping the movements of the boy, a kind of supernatural body swapping—never rose to the level of the conscious. As much as he yearned to be an ecologist, the animals that followed him home from the forest always felt off-limits as a subject of study. Until he was asked about the Quiznos coyote.

On Tuesday, April 3, 2007, during an unseasonably warm spring day, and against the backdrop of the restaurant chain's impending financial obliteration, a coyote walked into one of the conglomerate's downtown Chicago outlets. "Sauntered," corrects Schell with a laugh. "He sauntered in, hopped into a beverage cooler, and sat there for forty-five minutes. And folks who were making and eating sandwiches, they all backed out because they didn't know what to do. The coyote just fell asleep." Like so many urban animal stories, it was an in-joke that locals passed back and forth as proof of the hilarity of city life. *Isn't this place crazy? Anything can happen here.*

But when a panel of the world's foremost ecologists at the University of Chicago mentioned the Quiznos coyote to Schell during his grad school interview, something that he had long ignored suddenly came into precise clarity. "That question completely rerouted me," Schell says. "It was like, 'So I can do research where I live?'" Instead of going to some faraway exotic land, Schell was told that the faculty at the department didn't just see merit in exploring urban animals but were keenly interested in understanding how synanthropes adapt to city life. A coyote hopping into a drinks cooler at a Quiznos didn't have to be a joke; it could be the gateway to chart our changing relationship with the natural world.

One faculty member at Schell's interview, Dr. Seth Magle, saw a kindred spirit in the young would-be ecologist and proposed they work together. "Seth had this really radical idea," says Schell. "He was like, 'Why don't we just do a camera trap grid in the city of Chicago? Let's not focus somewhere else—on the continent of Africa, or go to Asia—no. Let's do research right here.'" Schell was ecstatic. In that moment, it was as if a lifetime of memories had finally converged. "All of these signals," he says, "were telling me that I can do research on those places that I love the most. That felt way more authentic to me as a Black man from Los Angeles than to go somewhere far-flung. That was the seed, and it just grew from there."

Quickly, though, Schell realized that the different urban ecologies he had experienced in Altadena and Pasadena weren't reflected in the discipline. He found no interest among the urban ecologists he met in exploring how wealth segregated social classes, and how that might influence conflicts between synanthropes and humans across the urban matrix, despite this being a central aspect of life in the city. It was odd to him—disconcerting, even—that nobody was seriously considering the question. "It felt like an insult," he says. "Me and other students of color would be in these academic spaces, talking about these issues, and either getting blank stares or being completely written off. And I was tired of it—of having separate personal and professional lives." I ask Schell why he felt those blind spots were important to push back against. "Because at a base level," he says, "it determines how we do our science."

He explains: "Most of the urban ecology studies done until 2020 were all sampling in wealthier areas. And on top of that, biodiversity in less wealthy areas is thought to be relatively low. The issue, though, is that it's biased." I ask him to elaborate. "It's confounded by the bias of who is doing the observations. I don't see white people going to Crenshaw or to the hood to try to study biodiversity," he says. "But we know that animals are there—or at least we don't know if we don't observe them. But all of these biases have stopped people from actually doing their damn jobs." As he learned more about the limits of urban ecology, the more passionate he became about undoing the systematic errors that had been plaguing the field. "If you're going to start working in a city," he says, "you need to know what it's like to live in a city. If you don't live in the city and don't know these injustices, then you're doing the field a disservice."

"Even with all that in mind," I say, "I guess most people would still assume that the main drivers of coyote–human conflicts are habitat loss, sprawl, and climate change, which force coyotes into cities. So, what does that miss?"

"The biggest piece," Schell replies, "is the social component. Everything you just mentioned is ecological. But sociodemographics, or residential segregation, or homelessness—these are issues that we have found strongly predict the likelihood of coming into conflict with a coyote." Schell gives the example of having dogs off leash, which is a major social predictor of coyote attacks. But that's only one dimension of the issue. A neighborhood's socioeconomic status, Schell has found, is predictive of our relationship with the canid, with wealthier urban areas tending to have a greater number of human–coyote conflicts. That's long believed to be a result of the luxury effect, Schell says, because wealthier neighborhoods also have greater access to nature and higher biodiversity.

"Some people would hear that and think that being poor is protective against coyote–human conflict," I offer.

"No," Schell replies. "Because the other piece of this is that the folks that are in wealthier areas feel like they have more agency over nature—more control—so they make reports to authorities. The folks that are low income, they don't have that, for all manner of reasons. They may not have access to the internet. They may not know what phone number to call. They may not feel like their voice is being heard more generally because of legacies of oppression. Or they'll see the animal and they may not ever actually come into conflict." Schell tells me about a study led by one of his postdoctoral fellows that investigated reports of coyote–human conflicts and quickly discovered that "conflict" was a relative term. "In many instances," he says, "it was just that a person saw an animal in their neighborhood. In those wealthier areas, though, people tended to interpret a sighting as a conflict. Why? Because those folks feel like, *Oh, there's a dangerous animal in our midst. Somebody needs to handle it*." That's a different attitude than most people in lower-income settings, Schell contends, who are much less likely to report when their paths cross with an animal's.

Still, I admit some skepticism, so I ask Schell how he can be so sure that coyotes really are active in lower-income settings if most of the research shows the opposite. "Because we have cameras everywhere!" he says with a laugh. Over the past few years, Schell has established, along with research teams, a system of camera traps across multiple cities, including Los Angeles, Chicago, and San Francisco. "The coyotes are in these low-income environments. They just aren't being perceived as causing conflict," he says. "So you see how the social component can be more important than the ecological one." According to Schell, this means the long-standing belief that coyotes only spend time in wealthier neighborhoods with lots of green space is just plain wrong, tainted by the systematic bias among ecologists against researching poorer neighborhoods.

Schell and his colleagues put this hypothesis to the test in Los Angeles, his hometown. While he had previously shown that, on average, biodiversity increases with a neighborhood's income in line with the luxury effect, that didn't prove the same trend would apply to specific synanthropes like coyotes. So he and his team collared twenty of the canids living in Los Angeles and tracked them over two years. Schell's team then mapped their movements according to multiple neighborhood indexes, including the levels of pollution, income, and vegetation, and the intensity of development. When he reviewed the data, the coyotes had given him a gift. Schell's maps revealed that the city's coyotes were deeply influenced by its wealth, but not in the way most ecologists believed. Instead, their paths avoided wealthier areas in the city, and they spent most of their time in highly polluted and lower-income areas, which they toured extensively. For Schell, that makes a lot of sense, because poorer areas in Los Angeles are also less likely to have quality services like weekly garbage pickup. Piles of garbage attract coyotes, which can feed off the refuse and also prey on smaller garbage-loving synanthropes like rats. The rats, in turn, can attract cats, which coyotes also eat, giving coyotes three meals in one place:

garbage, scavengers, and smaller mesopredators lower down the food chain. No part of the city, Schell had proven, was devoid of synanthropic life. Nature was an equalizer.

■ ■ ■

Schell entered the field during the second wave of coyote synanthropization, which has involved something of a role reversal. Unlike the first wave, cities are now less likely to ensnare coyotes in an expanding matrix; instead, it's the animals themselves that appear to be actively moving from wild habitats into urban ones in response to climate change. "Droughts, for instance, continue to intensify," Schell points out, "which makes many animals look for more stable access to fresh, clean water. And the best place to find that is in human settlements." Even when water isn't a direct draw, it can still indirectly bring in predators like coyotes, which are simply following their noses to their next meal. "Most prey species are looking for water and coming closer to people to find it," says Schell. "So now these carnivores are coming down the hill to get their food."

With waves of animals finding niches in cities, Schell began to wonder how urban areas would influence their evolution, leading him into the burgeoning field of urban evolutionary biology. "Say a highway splits two different sides of a city and serves as a barrier to animals moving from habitat A to habitat B," he says. "Then we know that their population size is going to stay relatively small. If the population size stays small, then the animals are more likely to go extinct because there won't be an exchange of alleles." Alleles are different forms a specific gene might take; the more genetic variety there is across a population, the higher the diversity of alleles that can be passed to future generations, and the more resilient the species will be. I think of the Berlin pipistrelles, forced by the acoustic urban environment of the city to change their calls, which might just end up with them becoming their own species. What Schell is describing is the night-mare opposite of that scenario, when populations are so small and static that the pool of alleles becomes smaller and genetic diversity dwindles to dangerously low levels. "That can get a lot worse in cities," Schell says, "where you don't just have one road, you have many roads. You don't just have one habitat getting smaller, you have all habitats getting smaller." The worst-case scenario is that cities so divide populations from one another that their genetic diversity bottoms out completely. When that happens, they can no longer evolve fast enough to withstand climate change, new diseases, or new forms of predation. "Populations need to be able to take a hit and still keep pushing," Schell says. Genetic diversity is the surest way for them to do that.

The upshot is that even if animals seem to be thriving in a city, they may be trapped in a multigenerational tragedy—a dwindling supply of alleles—that will eventually cause them to die out completely. Ecologists call this "extinction debt," when a population's current situation has doomed them to future annihilation. It's a little like watching someone get shot in slow motion: The victim is alive, yes, but the circumstances of their death are all but assured. Ecologists measure extinction debt by tracking the time from an inciting incident to the moment a species finally goes extinct. Oddly, this period is called "relaxation time"—as if extinction were a process of clocking out from work and casually kicking back in a La-Z-Boy as you wait for the universe to obliterate you.

Extinction debt is generally measured on the geologic time scale, in which a hundred years is akin to a single human heartbeat across the entirety of Earth's 4.6 billion years of existence. Beyond that, though, we still know very little about how it plays out. The obvious challenge is that animals on the verge of extinction are rare, and many go extinct without us ever laying eyes on them—invisible to us now, invisible forever. The International Union for Conservation of Nature, which tracks the vulnerability of the world's flora and fauna, estimates that roughly forty-four thousand insect species have disappeared over the past six hundred years. The number that have actually been documented during this time? Sixty-one, a paltry 0.1 percent of the total.

Aside from the fact that it seems to take a very long time for an inciting incident to send a species careening toward its extinction, practically the only other thing that ecologists agree on is that there's a relationship between extinction debt and habitat size. One broad rule of thumb is that a 90 percent reduction in habitat size can cause half the species living there to go extinct. The minimal data we have on this phenomenon comes mostly from remote islands, places from which animals cannot flee when habitats disappear. This is what happened—and is happening—at the D'Entrecasteaux Shelf, a seven-thousand-square-mile area off the coast of Papua New Guinea in the South Pacific, most of which is submerged about a hundred meters under the ocean. The animals living on the D'Entrecasteaux Shelf are slow-motion victims of the last ice age. Ten thousand years ago, as the planet warmed, liquid water trapped in ice was released, causing the world's oceans to rise by four hundred feet to their current levels. Over time, that left only the highest peaks of the D'Entrecasteaux Shelf above the surface, which formed an archipelago of islands. One of those peaks became Fergusson Island, a landmass of about five hundred square miles that supports eighty-six native bird species. While records show that the entirety of the D'Entrecasteaux Shelf supported more than one hundred different bird species before it was flooded, islands similarly-sized to Fergusson Island have an ecological limit of only about fifty-nine bird species. That makes Fergusson Island a place in evolutionary motion: an island supporting twenty-seven more bird species (the difference between the eighty-six it hosts and the fifty-nine it is ecologically able to support) than should be possible given its size. Those twenty-seven bird species may be singing beautifully now, but make no mistake: They are doomed to extinction by unstoppable forces unleashed ten thousand years ago.

Far closer to home than the D'Entrecasteaux Shelf, extinction debt is playing out in the Santa Monica Mountains, which border the city of Los Angeles to the northwest, a short drive from Schell's hometown of Altadena. The tony community of Calabasas, an exclusive enclave where the ultra-famous go to hide, is home to another less fortunate resident: the North American mountain lion. Despite the vastness of their range, which can extend to two hundred miles, Los Angeles' mountain lions are subject to the same fragmentation and enclosure that have forced other urban animals, like the Lincoln coyote and its brethren, to limit their movement. Unlike urban coyotes, though, mountain lions are poorly adapted synanthropes. As apex predators that have evolved to master forest habitats, they do not coexist easily with humans, while their reliance on big game like deer means that urban areas have little to sustain them.

As Schell says, animals need to be able to take a hit and still keep pushing through a change. In the case of Los Angeles' mountain lions, that change can be traced back to 1926 and the creation of U.S. 101, a major highway stretching from Southern California, up along the Pacific Coast, and all the way to Seattle, Washington. This paragon of modern American transport infrastructure passes through the Santa Monica Mountains, effectively dividing the sprawling forest reserve in two, with the southern portion bound by high-speed traffic along the freeway's six lanes, creating a nearly impenetrable wall of hurtling metal. It has transformed the Santa Monica Mountains into the modern equivalent of Fergusson Island: a habitat that has imposed extinction debt on the animals that live there.

And it seems to be unfolding within a blink of geological time. Trapped south of the 101, this population of mountain lions, estimated at about one hundred, has lost access to new alleles. Without them, the lions can't maintain a functioning and diverse genetic lineage, forcing them to trade a smaller set of genes to each subsequent generation until they can no longer survive. Some signs suggest this is already happening: A recent study found that 93 percent of male mountain lions in the area have a genetic anomaly leading to abnormal sperm, reducing their ability to reproduce. What's more, the inbred cats have begun to develop kinked tails, a sign that their genetic fitness is crumbling.

▪ ▪ ▪

At first glance, coyotes appear as unlikely as mountain lions to survive in cities. They're conspicuous and lack the cognitive flexibility that has allowed smaller mesopredators, like raccoons, to adapt to the dizzying number of distinct microhabitats that make up a cityscape. Their paws are brutish, like mountain lions. They're also a *K*-selected species, meaning they tend to have small litters of helpless young that require dedicated parenting, which is difficult to do well in unfamiliar environments (just ask the parents of the screaming baby on your next flight).* The most surprising thing of all, though, is that coyotes are deeply neophobic, meaning they have an innate fear of new things. And it is this last factor that should all but decimate their ability to adapt to the blinding novelty of the city.

Coyotes come by their neophobia honestly, and there's a simple reason for that: We fear them. The coyote has been widely denigrated for centuries across North America by colonists seeking to command and subjugate the land. "The coyote," wrote Mark Twain in his 1872 book *Roughing It,* "is a long, slim, sick and sorry looking skeleton with a gray wolf skin stretched over it, a tolerable bushy tail that forever sags down with a despairing expression of forsakenness and misery, a furtive and evil eye, and a long, sharp face, with slightly lifted lip and exposed teeth. He has a general slinking expression all over. The coyote is a living, breathing allegory of Want." While Los Angeles has been a coyote habitat for only about two decades, the hatred Twain

---

* That's in contrast to *r*-selected species, like most insects and the giant Pacific octopus, which produce high-volume broods that they quickly abandon.

preached is as active today as it was in the nineteenth century. *Have you lost a pet to one of these Feral Doggies?* reads a typical post from a Los Angeles–based user on the Nextdoor app, a hyperlocal social networking platform for neighborhood residents. *Have you witnessed your pet being carried away after her spine has been snapped? As she uttered her last bone chilling warrior screech? Followed by her final grunt as she relinquished her power to the jaws that carried her away?* No wonder the coyote is inherently neophobic; it has long been considered a supernatural monster with the power to undo the static social order that humans have long cultivated. And in an age of viral media, each new account of a coyote attack (or "conflict") can intensify the human fear of coyotes, and in turn further the canids' terror of us.

There is some irony that the recent movement of urban coyotes into massive cities like Los Angeles and Chicago has led us back to an age-old, pre-colonial, and quasi-mystical view of the animal's powers. Long before European settlers arrived, the coyote was worshipped by Indigenous nations across the continent, from the Aztec in central Mexico, the Apache, Hopi, and Navajo in the deserts of the U.S. Southwest, the Lakota and Cheyenne of the plains, and even the Syilx Okanagan and Secwepemc peoples in what is now British Columbia, Canada. In all cases, Coyote is a multifaceted god: selfish and benevolent by turns, a Promethean fire stealer who shares secret knowledge, a creator of humans, and a willing teacher of dance, hunting, mischief, and song. In one story shared by the Maidu, who reside in what is now Northern California, all the animals compete in creating humans in their own image by making sculptures. Coyote, the night before a winner of the contest is to be chosen, smashes all the other sculptures, leaving only his own intact. This is why humans, the story goes, are as sneaky and wily as Coyote himself. More than anything, the coyote has long been revered as a shape-shifter, capable of survival and resurrection after death, which he accomplishes by blending into the background and adapting to whatever situation confronts him. These are ancient stories from ancient cultures, some dating back ten thousand years or more, making Coyote the oldest continuously worshipped deity in North America. And with the rise of the urban coyote, the extent of the truths these myths hold is coming into focus once again.

The story of Coyote sculpting man in his own image is instructive if you want to understand how a deeply neophobic creature became the largest successful synanthrope in North America. One study of captive coyotes in Utah yields some fascinating clues about how well the canids are primed for human environments. Scientists set up a series of six pens that mimicked natural and urban environments in different proportions, ranging from an environment that included only native grasses, trees, and shrubs ("100 percent natural") to a pen that contained only plywood and wood pallets, trash cans, a culvert, and solar lights ("100 percent urban"), with other pens containing varying mixes of the two in between. They then let the coyotes choose which pens to den in over a period of four days and found, remarkably, that the animals spent two-thirds of their time in the pens that contained urban materials, with 29 percent of their time spent in the two pens that mostly or entirely mimicked a city environment. What was most surprising was that none of the captive coyotes had ever seen, let alone lived, in urban areas. Even so, there was something about the artificial materials that called to them. Deep in the coyote soul, the experiment suggested, lies a love of the city. What's more, when the scientists analyzed the coyotes' choice of pen by their personality type, they found that bold coyotes spent 85 percent more of their time in "urban" pens compared to non-urban ones, which was four times as often as the shy ones did. The experiment raised the striking possibility that the materials we use to construct our cities attract the boldest among them.

Chris Schell has studied this phenomenon to try to make sense of how coyotes become bold, and how this might be shaped by human interactions. Because coyotes are *K*-selected, their worldviews are deeply shaped (as are our own) by the outsized influence of their mothers. Schell was curious to understand whether a mother's experiences with humans would influence the personalities of her offspring.

In one experiment he ran with captive coyotes, mothers were observed over two successive litters; in both, between five and fifteen weeks after the birth of her pups, a human observer sat near the coyotes' daily food source to mimic the experience of living cheek by jowl with our species. After the birth of their first litters, the mothers were terrified of the observer and kept their distance from the food. After the birth of their second litters, though, that fear had waned and they nonchalantly passed by the same human on the way to their food. What's more, that personality shift was picked up by the pups. The first litter's whelps were far more skittish than those of the second, who were born at a time when their parents no longer feared human interaction. This explains, says Schell, why "each generation is bolder than the last."

But that boldness can be a death sentence for coyotes that return to rural areas. When an urban coyote comes into conflict with humans, the solution is most often relocation. In Colorado, for example, nine coyotes are removed each year from the metro Denver area, with euthanasia used only as a last resort. In rural areas, though, the U.S. Department of Agriculture's Wildlife Services division (sometimes dubbed the "killing arm" of the USDA) regularly carries out mass eradication programs that target entire local populations with M-44 cyanide bombs, neck snares, foothold traps, and aerial gunning, killing about seventy thousand wild or semi-wild coyotes every year (for context, the next most frequent target of Wildlife Services are rural foxes, of which two thousand or so are killed annually). This creates a feedback loop: Boldness in rural areas is punished severely, making it an undesirable trait for a coyote looking to stay alive. In the city, boldness won't get a coyote shot (or bombed), but it can reap them all kinds of rewards, like better hiding spots, more reliable access to water, and proximity to densely packed prey that they can feast upon. And this is doubly the case when the sun goes down. Like many synanthropes, coyotes in cities have become increasingly nocturnal over the past few decades, a recognition that you can get away with a lot more in a city under the cover of darkness.

Schell is loath to directly compare himself to the animals he studies but admits that he feels a kinship with the way coyotes have adapted themselves to unfriendly territory. "I liken coyotes to the experience of being Black in America because they're hyperpolarizing," he says. "They're hated on. They're feared. People try to eradicate them from an environment. They don't leave. They learn quickly and figure out ways to survive in cities." When he began to track coyotes, often in the dead of night, Schell also had to quickly learn how to manage situations that might end in conflict. "I'm not scared of coyotes but cop sirens when I have an anesthetized animal at three A.M.—that'll raise my hackles." To preemptively de-escalate situations, he drapes himself in scientific emblems. "I wear the nerdiest glasses I have and a jacket that has my college logo, so that people don't mistake me for what they think is a thug." That, he says, "gives me some credibility beyond just being a Black person out in nature."

▪   ▪   ▪

At first blush, pigeons—derided for their stupidity, monochromatic design, and overabundance—may seem wholly unlike the awe-inducing, shape-shifting, and frequently solitary coyote. But the two animals share one important trait: a dulling of their flight response in the company of humans. Many urban animals who thrive in cities do so because they have fewer predators there, so they slowly get desensitized to creatures that might, in another habitat, pose a threat. While they may react to humans at first, the high density of us in cities makes it impossible for synanthropes to react with the level of wariness and alarm they would otherwise muster in natural spaces. Doing so would be too energetically exhausting, and they wouldn't be able to consume enough calories to survive. That is as true for pigeons as it is for coyotes, as well as practically every other synanthrope with whom we share space.

Thick flocks of pigeons pecking at cigarette butts on a downtown street, as if the gray and mottled sidewalk itself had turned into some kind of viscous flapping substance, feel like an accident of urban design. Their plodding gait and their slow reaction time to our footsteps as we march through their numbers suggest a species remarkable only for its stupidity and laziness. That they are exactly in our path feels less intentional and more like the actions of a barely animate creature, like piles of fallen leaves clogging up storm drains after a rainfall. They are as conscious, it seems, as a discarded coffee cup. But that read, according to Schell, misses one of the most subtle and effective adaptations that pigeons and other synanthropes have made.

Both coyotes and pigeons, Schell says, haven't just dampened their reaction time to humans. They've actually learned, through trial and error, to override their inherent biological responses to predation and actively run toward danger. That has meant embracing *Homo sapiens,* despite the long and murderous history of our species. "Embrace" might seem hyperbolic, but it's not quite a metaphor. "Pigeons in cities," Schell explains, "go where pedestrian traffic is higher, because being around people provides what we call a human shield." It's a seemingly counterintuitive move until we expand our minds to think not just about our relationships with other synanthropic species but the relationships they have with one another.

Human hostility isn't doled out equally. Instead, we save our most intense wrath for predators that occupy similar niches to us. We'll shoot a bear that gets too close, but we'll shoo a pigeon that climbs onto our boot. Those differing reactions become common knowledge across the animal kingdom, even among animals we might consider dimwitted. "Many prey species sort of know that," says Schell, "so then they get closer to people to guard against predation." To hear Schell tell it, pigeons haven't wandered into our path by accident: The birds have come to realize—even if they don't realize they're realizing it—that there's a negative linear relationship between the number of humans in one place and the number of other predators hanging around. The passive-aggressive derision we send their way is nothing compared to being devoured.

Schell's human shield theory elegantly explains why pigeons glom on to humans like microplastics. At face value, though, it doesn't explain why coyotes have been moving with such ease into megacities like Los Angeles. If anything, the human shield theory should mean that coyotes would be less likely to follow prey into habitats packed with humans. That's true, until you consider the mass eradication campaigns carried out in rural areas by the human staff of the USDA. Simply put, rural coyotes are much more likely to be shot (or bombed or trapped by the neck or strafed from the air) and killed, while urban ones are generally tolerated. And this has set up a weirdly self-referential situation: Coyotes have entered cities to flee human predators and, in so doing, are using urban humans—weak-hearted, blithely mild predators—as shields. With every deadly cull that happens in the countryside, coyotes are incentivized to enter cities. With every toss of a half-eaten donut to a patient coyote in a city park, and every stone-still amateur photographer snapping pics instead of pulling a gun, urban coyotes absorb the same message: *It's okay to let your guard down.*

"Just to make sure I fully understand," I say to Schell, "there are two theories. The first is that pigeons no longer react to humans in cities because there's too many of us, so their response to all predation—not just from humans but from all animals, just by being around us so much—goes down."

Schell nods.

"But the other theory is that urban animals do react to other predators, but not to us. And these prey species are attracted to us because humans in cities act as a buffer against other predators. Is that right?"

"Yeah," says Schell. "And here's the cool thing: There's this cognitive arms race. It's not always the case that predators just don't want to be around people and that's it. When resources get tight, you've got to make it work. You've got babies to feed, right? So, there have been species of raptors that have figured out how to exploit cities too. And when they get into a city, the anti-predator response starts to go back up in pigeons, up to non-urban levels. Why? Because all of a sudden you have a peregrine falcon or a red-tailed hawk that's, like, 'Let me just snack on these birds real quick. They're not ready for me, so I'm just going to take out twenty at a time.'" Schell finds this behavior change so fascinating because it demonstrates that even animals like pigeons and coyotes that appear to be cognitively fixed (unlike the flexible mind of the raccoon) can, if conditions change, recall lessons from their past to survive whatever possible futures may arrive.

■ ■ ■

And conditions can change in a heartbeat.

It's January 15, 2025, and I'm in Los Angeles. All around me, multiple wildfires are raging. The Palisades Fire, which started a week ago in the Santa Monica Mountains, is still burning out of control, with roughly 24,000 acres destroyed. So is the Eaton Fire, which ignited a few hours later in the foothills of the San Gabriel Mountains, and which has in just a few days burned to the ground vast swaths of Chris Schell's hometown of Altadena, part of a torched landscape that stretches across more than 14,000 acres. The devastation is shocking. People living tens of miles away have found a light dusting of ash, ominous and as fine as snowflakes, covering their porches, barbecues, lemon trees, and electric cars. Ten days after their ignition, I can still taste the bitterness in the air, a sharp tang in the back of my throat, the city transformed into particulate.

Schell's family members were among those who lost everything. His uncle and aunt saw their home reduced to ash, the only remnants a charred set of barbells standing amid the smoking wreckage. His grandmother's house also burned to the ground, which was particularly painful for Schell. Her house, he says, "was my home, in a deep way. It was a safe place to be ourselves, a hub for family, a place that if everything went to hell in a handbasket, we could find refuge there. It was the physical manifestation of love and community, and it rocked our core when it was gone. It was like getting punched in the chest." His family's homes in Altadena stretched back four generations, to the Second Great Migration, which saw more than 5 million Black

Americans exit the South for safer places across the United States. All in all, the Eaton Fire forced his mother, grandparents, brother, aunt, and uncle to flee their homes and the lives they had built in Altadena. "Becoming a statistic after all of the papers and books I've read around climate change and environmental injustices," says Schell, "to then now be a part of it, is especially painful."

Beyond the human toll (150,000 displaced and 29 killed), the wildfires scattered countless animals, including ground squirrels, deer, owls, bobcats, coyotes, mountain lions, and even black bears that had been living in the city and its adjoining forests. Wildfires will burn through forests unfettered; once they pose a threat to human communities, though, resources will be ratcheted up to halt their trajectory. From an ecological perspective, putting out fires is just another way that urban areas attract wild animals.

While Los Angeles is still battling to contain the blazes and grimly accounting for the scale of the destruction, climate refugees have begun showing up in neighborhoods spared destruction. One coyote, disturbingly thin and wearing an arrangement of burn marks on its fur, dragged itself through Altadena at midday, desperately searching front yards and back gardens for water. Across town, on Palisades Drive, a coyote, its fur charred to a thin layer and its normally bushy tail reduced to a stick-thin appendage drooping like a dangling worm, wandered in the middle of the four-lane byway, stumbling and disoriented, the mountains, burned to black, rising up behind it.

They've also been seen, I'm told, wandering a neighborhood called the Oaks. It's a tony enclave high up in the hills next to Griffith Park, a four-thousand-acre green space that sees roughly 10 million human visitors every year wandering through its scrubby mountainous hills, along with coyotes, mountain lions, bobcats, and mule deer. In recent years, Griffith Park has become known as the home range of P-22, a mountain lion that during the COVID-19 pandemic briefly became the most famous synanthrope in North America before being found half-dead in a nearby yard, the victim of a car collision, and subsequently euthanized. Normally, coyotes flock to Griffith Park every morning because of the endless food waste left there by its twenty-seven thousand daily human visitors. But authorities have decided to close the park during the fires, leaving the local coyotes with nothing to scavenge. Set atop succulent-covered canyons adjoining Griffith Park, the Oaks is the platonic ideal of Los Angeles neighborhoods, with sprawling Spanish Colonial Revival and Craftsman homes barely hanging on to the surrounding cliff faces, while winding roads, the perfect backdrop to a Lynchian psychodrama, go dark at night, becoming unlit corridors that dissuade human foot traffic and attract shy nocturnal creatures.

I have been sitting for three hours at an intersection in the Oaks where five squiggly roads come together, watching the sun set, the dusk blanket the sky, and night set in. The foot and car traffic of the late afternoon is gone and nobody is outdoors except for me. I'm feeling sorely out of place, deep in the silence, while seated on the edge of a short retaining wall in front of a large empty corner house. This isn't somewhere people sit in the dark for no reason, but after chasing raccoons, octopuses, and bats across cities, I've learned that you need to change your rhythms if you want to commune with urban animals. The anxious and very human feeling of being at the wrong place at the wrong time usually means you're exactly where you should be.

Then I hear them. A pack, at least a dozen voices, howling recklessly as they move along the top of a cliff behind me. Each call starts as a high bark, then transforms into a long and sustained howl, building momentum until it is joined by a chorus of cries that sound to me like unbridled freedom. I stay still and silent, but they are moving, the howls now radiating to my left, the pack arcing themselves across the canyons, liberated by the retreat of humans at night. I can't help but feel relief: Their lusty calls are the first exclamations of joy that I have experienced since the fires began. The howls then shift again, coming from the area in front of me but quickly fading, the coyotes following familiar footpaths away and into the wilderness of Griffith Park.

I get up and wander around in a tight circle to clear my head, the incantations still reverberating in the darkness and the silence that has folded itself over this place. Suddenly some sense fixes me, and I turn to look up the poorly illuminated road behind me. There, about twenty feet away in the middle of the sloping street, is a solitary coyote, staring down from above as it sizes up the scene. Its light brown fur is thick and glossy in the dim lamplight and it radiates absolute calm. I pause, dead in my tracks. It pauses too, just for a moment, before continuing down the hilly road toward me, searching for something in the grounds of this empty house outside of which we are both loitering. For reasons of its own, it has boldly decided to leave its pack and wander through the night alone. I watch, stone-still, as it carries out its investigation, its muzzle trained on the house's surrounding hedgerow, until it's only about fifteen feet away. I fidget, wanting to move closer, but am unsure of what to do. The coyote stops again and looks at me, sizing up my frantic energy and finding it wanting. Unhurriedly, it turns and lopes gracefully back up the street and disappears into the night. It leaves me with something I have felt so many times before but have never been able to name until now: the feeling of not having howled.

▪ ▪ ▪

As soon as they enter urban areas, coyotes and other mesopredators tend to reduce the size of their home ranges substantially. It's a product of the attraction of predictable supplies of nutrients (garbage) and shelter (roofs, decks, and culverts) and a lack of paths leading to other, better habitats. Once ensconced in cities, synanthropes find it hard to leave.

This might give the illusion that an urban species is thriving when it's in fact mired in an "ecological trap"—a low-quality habitat that is more attractive than other places where it would have a better chance of survival. Ecological traps are exceedingly rare in natural environments, but more than thirty different types of traps have been identified across the labyrinthine architecture of urban spaces. One ubiquitous trap is the fast-moving freeway continually depositing animal carcasses on the side of the road after collisions. The carrion landscape it creates attracts coyotes and other urban scavengers, like rats, crows, and vultures, who feed well and often but are also themselves more likely to be hit and killed—there's the trap—becoming the next item on the menu. Thick flocks of gulls scarfing discarded food at a seaside market are the epitome of successful synanthropes, but that is just the bait. The birds appear safe and well-fed, but the dreck they eat—ice cream, battered fish, french fries, and ketchup packets—is killing them, unlike the nutrient-packed fresh fish, mollusks, and insects that will keep them healthy and alive. But why hunt when you're literally surrounded by "food"? The birds can't help themselves, becoming undernourished, weak, and vulnerable to disease and death—a slow-motion trap as cruel and effective as glue traps for mice.

In some cases, the trap is embedded in the materials we use to build cities. Some forms of asphalt polarize light horizontally, making the sun's rays stretch wide across an animal's field of vision. In mayflies, dainty aquatic insects that require water sources to proliferate, sunlight on asphalt creates a "supernormal stimulus" that looks to them more like light bouncing off a large pond than the real thing ever could. Mayflies swarm toward the trap and end up smeared across your windshield in a cycle of ecological degradation that has profound implications for the sustainability of bats and birds that feed on the insects; it's a trap that draws more animals in the longer it unwinds. Indigo buntings, startlingly colorful songbirds painted in brilliant shades of blue, are attracted to forest edges. That's because these liminal spaces are generally bordered by early successional habitats—environments in rapid transition, like a scorched piece of land after a wildfire—and generally harbor fewer predators. Cities, early successional habitats fixed in time forever, upend that calculus. When forest patches are set against the hard edge of an urban space, indigo bunting biology cues the little birds that this is a safe place to lay eggs. But urban areas, as we've learned, are where the coyotes are, along with red-tailed hawks and peregrine falcons, often in numbers far higher than in their natural habitats. And so, after indigo buntings establish their nests, coyotes are there to kill the birds and steal their eggs. The trap, laid perfectly, ends with scarlet stains on the brilliant blue mosaic of their tattered feathers.

▪　▪　▪

The story of the Lincoln coyote is the perfect cold open for the longer tale of the coyote's evolution into an urban mesopredator. That brief case report captures the key moments in the coyote's transformation: its enclosure within the matrix of an expanding city, its mixture of curiosity and wariness, the fear it elicits, its desire for companionship, and finally, its end.

A month after its last playdate with the golden retriever, nine days before Christmas 1976, the Lincoln coyote was shot and killed by a hunter as it roamed a rural habitat five miles outside of its home range. Despite having found an uneasy balance with the humans with whom it shared the city, it was, ironically, the coyote's newfound boldness that led to its death once it ventured back out into wilder places. We can never know why it traveled so far from the sorghum fields and flirty canines that made up its quotidian routine. I like to think that it was because of lost love, which can make you do crazy and self-destructive things. Whatever the case, the coyote had been profoundly transformed. When it entered Lincoln for the first time, it did so as a wild animal. When it left a year later for what would be the last time, it had been stripped of its neophobia. The final ingenious trap set by the city was in its own mind, returning the coyote to the natural habitat of its forebears as a guileless creature, made foolishly soft, and marked for an easy death.

Part II

Damnatio ad Bestias

The Roman spectacle of feeding prisoners to lions and bears, known as "condemnation to beasts," is an ugly but revealing chapter in human history. At the peak of the Roman Empire, around 100 C.E., wide swaths of Eurasia were populated by wild animals, leaving ordinary citizens at risk of mutilation or death if they left cities and ventured out into forests alone. With ferocious beasts pressing at the gates, watching nature's cruelty play out within the safety of a purpose-built arena was a triumph of man's separation from nature. No longer would the blank ruthlessness of natural laws decide who lived and died. Instead, the citizens of Rome could play the role of omniscient gods.

At least, that was the case within the circular walls of the Colosseum. In the rest of the realm, Romans, for all their supposed power, were unable to stop the encroachment of wild animals. The cities and towns of the Roman Empire were filled with familiar synanthropes that had found generalist niches and had been borne across vast distances by the carefully constructed roadways and shipping routes fanning out from Rome across the many lands of the

Empire. It was the Romans, it turns out, who spread *Mus musculus,* the house mouse, across Europe and the Near East; it was the Romans too who were eventually beset with *Rattus rattus,* the black rat, which burrowed its way into the capital's grain supply and found its home within the walls of Roman apartment housing blocks, known as *insula,* refusing to ever leave.

Rome has many legacies, among them its relentless standardization of human environments. Roman urban design remains the foundation of many of Eurasia's greatest cities and many more across the world. It's why, even in cities far from home, we're likely to encounter roads set out on a rectangular grid, with a core across which two main streets intersect, a downtown forum replete with civic buildings and shops, and a city design split into separate blocks. It's also why, two thousand years later, homes and monuments around the world are still built with the same basic materials used by the Romans: brick, stone, wood, concrete, glass, and iron.

The Romans saw order and organization in their cities, and for the most part, so do we. Urban ecologists see something else: homogenization, standardization, and the replication of identical habitats, no matter where on the planet they are located. And it's getting worse. In a study of 150 of the largest cities in China, India, and the United States carried out between 1995 and 2015, scientists found that cities were becoming increasingly homogenous in their shape, their size, their sprawl, and the connectivity of their internal fragments—those microhabitats that raccoons, coyotes, and other animals must learn and unlearn to navigate. What's more, the fastest-growing cities on the planet are the most alike, portending a future in which more people live in identical environments.

The famed ecologist Charles Elton once likened Earth's continents to large liquid-filled tanks, each with their own chemical solution, connected by tubes that were blocked by closed taps. Humans, he explained, have been slowly opening the taps and letting the different liquids mingle, and we have been opening them more quickly as time goes by. Eventually, the liquids—which in Elton's analogy represent the unique and biodiverse ecosystems on Earth—will all merge into one homogenous fluid as long as we continue to build the same urban systems with the same connective links between them.

As I noted in chapter 4, ecologists understand cities as early successional habitats, immature environments in the midst of a trajectory, like young forests and field edges, that are most often created through some massive disturbance. Early successional habitats aren't devoid of life: They are where grasses grow, along with small shrubs, herbaceous forbs, weeds, and flowering plants that thrive in direct sunlight (like wisteria, a purple-flowering vine that often grows in the wake of forest fires). They also tend to attract specific animals that do well in these tumultuous environments, like high wing loading bat species adapted for open flight, and *r*-selected species like insects and small rodents that generate and abandon high-volume broods. That makes *r*-selected species well suited, of course, to prospering in cities.

In nature, early successional habitats are shaped through different forces (fire, flood, avalanche, hurricane), but they all have one thing in common: They are temporary. One of the most peculiar aspects of cities, then, is that they are early successional habitats that have been frozen in time. The chaos of urban ecologies are, from an ecological perspective, meant to be short-lived, inevitably on their way to becoming something grander. Humans, though, have halted their trajectories and made them permanent fixtures, humble and homogenous, placing cities in a state of disequilibrium with their surrounding environments, out of sync with the natural world. This ends up repelling many species while inviting the same ones in regardless of where the city is located on the globe, creating what ecologists describe as "biotic homogenization": a dwindling diversity of resident species.

This has led to a situation in which humans experience new cities as exotic and distinct, but synanthropes experience them all as essentially the same. And they have for millennia. As far back as Greco-Roman antiquity, we have been paving the way for synanthropic invasions to happen in uncannily similar ways across cities divided by continent and culture. If *Mus musculus* can invade Rome, it can invade (and has) Vladivostok, Tehran, Dar es Salaam, and Buenos Aires. And so can *Blattella germanica* (the common German cockroach), *Passer domesticus* (the house sparrow), *Rattus norvegicus* (the brown rat), dozens of Chiroptera (bat) species, and many more. The predictability of urban design functions as a meta-communicative signal for synanthropes to enter.

Most of the time, we barely notice synanthropes; sometimes, we tolerate them. But there are those rare urban species that refuse to be ignored. These are a class of invaders that haven't found stable niches in cities but instead foment instability across the social order. And it is their rapacious thirst to exploit, dominate, and control that risks breeding chaos in the very places that we have so long seen as refuges from nature's chaos. We have set the table for their conquest and now watch helplessly as they dominate landscapes we thought we ruled.

Chapter 5

Neophytes

And so the one in our garden continued its growth peacefully,
as did thousands like it in neglected spots all over the world.

It was some little time later that the first one
picked up its roots and walked.

—John Wyndham, *The Day of the Triffids*

Invasions Begin with Invitations. The most successful synanthropes find ways to thrive in cities, often against all odds. In some cases, the success of a synanthropic species is so outsized that they're able to not only find sustainable niches but also actively reshape the urban environment itself. These are synanthropic invasions, which can be terrifying but rarely occur without a helping hand. In most cases, unstoppable invasions happen when humans introduce new species into landscapes that cannot withstand them. And that's as true of urban animals as of floral invaders. Plants—flora—inhabit our cities, surrounding us while growing, reaching, unfurling, and sometimes destroying. And while it's understandable to think of them as a part of the landscape rather than as active participants in it, synanthropic plants are just as capable of unbalancing ecosystems as invasive animals. What's more, it is plants, not animals, that are overall better equipped to not only invade urban environments but also maintain their dominance within them. We can contain raccoons, relocate urban coyotes, and retrofit buildings to stop bats from entering them. But will we ever get rid of the dandelion?

∎ ∎ ∎

"While She's Gone," the headline blared, "'Burglar' Enters." Buried on page 10 of the August 20, 1952, edition of *The Birmingham News*, amid a sea of classifieds advertising Mexsana Powder Shower ("GREAT FOR HEAT RASH") and Marvin Hawkins's Pen Repair Shop ("Just Like at the Factory"), a short article recounted, in breathless prose, the kind of grisly crime that tantalized readers looking for a jump scare over their morning coffee. An elderly woman. An empty house. A ruthless invader. A home destroyed. It was the kind of piece that left most readers with a mild frisson before turning back to their breakfast and thinking, *Sure, but it couldn't happen to me.* Except that here, in Birmingham, Alabama, and now, in 1952, the stakes were much higher. As they scanned the piece, the daily's loyal readers would have all felt the same rippling anxiety: *It's happening to everyone.* The home invasion wasn't an isolated incident. No. Their city—pride of the Deep South, a bustling metropolis built on steel and the industry of man—was being forced to its knees by a wholly inhuman power. The worst of it? Not so long ago the invader had been welcomed in with open arms.

"Mrs. Jo Farmer's living room had a jungle motif she didn't plan on when she returned home yesterday from a month on Cape Cod," read the piece. "The huge kudzu vine which almost covers her home found its way through a slit near a window and grew into the living room. She found long tentacles entwining picture frames, the piano and other pieces of furniture."

It might seem hyperbolic to describe a plant as prehensile until you learn that the kudzu vine can grow a foot a day. In a very real way, it can—and does—move, dragging itself swiftly across the land like water running downhill. When it washes over landscapes, it immerses them in a dark glaucous sea, leaving houses, telephone poles, copses of trees, and any other structures in its path looking like vague sunken shapes forgotten beyond the abyssal layer.

One of kudzu's most effective defenses from eradication is the sheer size of its bulbous rhizomes, which can grow to three feet or more under the ground, making it almost impossible to uproot, dislodge, or kill the plant; some unconfirmed reports claim that some rhizomes weigh up to four hundred pounds. Meanwhile, on the surface, each shoot the vine sends out will, once it finds dirt, establish a new rhizome, spreading the matrix ever farther. Within a year, that new crown will be independent of its progenitor, providing kudzu with a new base of operations and a new head to lop off, marking the birth of a totally decentralized organism. When rhizomes dot a landscape, there is no easy way to kill the mass. For proof, all you need to do is visit Birmingham, Alabama, where the burglar that slipped through Mrs. Farmer's living room window in 1952 still roams unchecked.

▪   ▪   ▪

Birds chirp outside the Birmingham-Shuttlesworth Airport, sounding like car alarms going off in the well-manicured trees. In East Birmingham, at the edge of the urban boundary layer, tarps covering the cylindrical hulls of rusted smokers flap next to dilapidated one-story houses, all set under a mottled gray sky. In downtown Birmingham, conglomerates of skyscrapers are set far apart from one another, looking like boulders tossed together, the work of central planners who expected a never-to-arrive influx of people and capital to fill the in-between spaces. Instead, the clumped and negatively spaced topography throws up curious oddities, like the headquarters of the Jefferson County Department of Health, a squat five-story building dating to 1979 that rises like a terraced Aztec temple out of the surrounding emptiness. Due north across town, the AT&T Microwave Tower, a 1960s-era relic of the wireless communication age, sticks up haphazardly from the roof of another flat high-rise like a toddler's perilously placed wooden block, a lonely urban cliff glowing in the night, the circular aquamarine tint of the corporate logo seeping light pollution across the sky.

Birmingham is a town like any other, which is to say monochromatic and predictably built but harboring its own disquieting troubles. It's not that hard to seek them out: No matter which neighborhood you start in, all you have to do is follow the vines, and you'll soon find yourself entangled in their undiscriminating power. Kudzu is draped all over, covering up hedges and drowning vacant lots in its dark, almost blue-black verdancy. Once you see it, you can't look away: kudzu, working its way up tree trunks, forcing branches to hang pendulously, making everything here—the trees, the power lines, gutters, and shrubs—bend and bow, like supplicants before a king. Birmingham isn't the only city grappling with kudzu. The perambulating vine is often said to cover an astonishing 7 million hectares of U.S. land across 27 states, and to spread at a rate of almost 124,000 acres per year. The Deep South—Mississippi, Georgia, and Alabama—is the epicenter of the plague, though kudzu has been creeping ever outward as the world warms and is now found as far south as Florida, west through Texas and Oklahoma, and even as far north as the Canadian province of Ontario.

But Birmingham is its patient zero. It's the only city where I have seen a landscape being choked to death before my very eyes. Its biotic homogenization of the South is doubly disturbing, because this region is a refuge for species that retreated from the glaciers that spread across the continent 18,000 years ago. Today, the U.S. Southeast has the highest level of plant diversity in the country, with more than 160 endemic species, including many "ancient relicts"—the last surviving members of plants that were once found all across North America prior to the most recent ice age.

▪   ▪   ▪

Kudzu is a notoriously tenacious synanthrope, but it is only one among many with the power to overrun urban ecologies. Rabbits entered Australia 150 years ago and have since become an unstoppable scourge. The island continent's rabbit plague has been traced back to a single introduction: twenty-four bunnies imported in 1859 by an English settler named Thomas Austin, who wanted prey to hunt on his property at Barwon Park, near Melbourne in southern Australia. Those twenty-four were enough to seed a biological invasion (and proof that Austin was probably a pretty poor shot). By 1866, seven years later, Austin's hunting party killed more than 14,000 rabbits on his property, an exponential increase that set the stage for an environmental apocalypse. During the Great Depression, fifty years after Thomas Austin's strategic error, more than a million rabbits were killed and canned on the island continent, the meat sent to Australian servicemen fighting in World War II. Today, there are more than 200 million rabbits in Australia. A classic r-selected species with the ability to produce as many as seven litters per year, an absence of natural predators, and a massive landscape to move through, rabbits have expanded across Australia at a shocking rate of sixty-five miles per year and now cover roughly 70 percent of the continent's landmass. That has made it the fastest-known mammal invasion ever recorded, and one that costs the country the equivalent of US$130 million per year in lost productivity.

Over the past century, every attempt to stop the spread of rabbits has failed. These include, briefly, beating them to death with sticks (inefficient), a three-thousand-kilometer-long fence (the rabbits ran around it), the release of European foxes into the countryside (they hunted endangered marsupials instead), the introduction of a killer virus (the rabbits developed resistance), infection of a parasite via fleas (it infected native species), and cruder methods like air-dropping poison, bulldozing warrens, and even planting dynamite underground to blow up their colonies, colloquially known as fluffles. Thomas Austin's rabbits have survived them all, devouring native plants and disturbing the life cycles of native animals, and leaving about three hundred species at risk of extinction, including the pygmy possum, the orange-bellied parrot, and the ballerina orchid. All because one man assumed that they were his prey.

Human invitations to invasive species have taken many forms. In New Hampshire, the town of Grafton elected a libertarian caucus that effectively dissolved basic government services—including garbage removal—leading Grafton to become overrun by black bears. In New Delhi, macaques are revered at the temple of the Hindu monkey god Hanuman, where they are afforded the same religious reverence as their mythic protector and have taken full advantage, spilling out through every crevice and bullying their way to tourist provisions, no matter whether offered or stolen. Churchill, Manitoba, a sleepy Canadian town of nine hundred people just south of the Arctic Circle, has long been known as "the polar bear capital of the world," as it sits along a key corridor the bears use to wait for sea ice to form. But as melting glaciers have led to more and hungrier bears roaming through arctic towns, that whimsical slogan has become something of a distress call. From just a handful of bears a few years ago, there are now as many as seventy-five polar bears each year marauding through Churchill, and the numbers are rising.

■   ■   ■

Kudzu doesn't have long teeth, claws, or floppy ears, but it is an exemplar of the long-standing hyperefficiency of synanthropic plants. Look around and you will no doubt find others lining the cracks and patches of your neighborhood. Dandelions, bull thistle (a spikey purple-flowering plant), broadleaf plantain (a large-leafed plant that sends up a single long and thin petiole from the center of its rosette), goutweed (a small shrubby weed in the carrot family with white-tipped leaves), and even Kentucky bluegrass (a.k.a., your lawn, with sales cresting $100 million annually) make up the greenery of many cities. It's nothing new. Archaeologists have found many contemporary weeds at a hunter-gatherer site on the Sea of Galilee dating to 23,000 years ago.

Despite that long history of coevolution, it wasn't until 1912 that the first-ever census of floral synanthropes was attempted, by Swiss botanist Albert Thellung. Thellung was a diligent but chronically ill young man who had long been obsessed with parsing out the various plant communities that covered Western Europe. His obsession with invasives would find its crowning achievement in his book *La flore adventice de Montpellier,* which set out to record every single synanthropic plant growing in the southern French town of Montpellier. Seven years and 688 pages later, Thellung had catalogued 800 invasive plants, hundreds of which he'd traced back to the seventeenth century, when the region was a major processing hub for the international wool trade. Thellung's eye-popping systematic rigor transformed the entire field of botany.

Thellung proposed that there were three phases of every biological invasion, known as "naturalization" stages. The first is when an alien species has been introduced into a new environment but isn't yet able to reproduce. The second is when an alien can propagate itself in the new environment but could also be easily wiped out if a small apocalypse—fire, flood, drought, or exerted eradication by humans— arrived. And the third stage occurs when a species has fixed itself in place, permanently, and is able to survive and replicate over multiple seasons without human intervention. Instead of describing this process as an invasion, Thellung called it "*Pflanzenwanderungen unter dem Einfluss des Menschen,*" which translates as "plant migrations under human influence," a more apt description for a process that so often starts with human error.

Kudzu (*Pueraria montana*) fits this mold perfectly. Like many other invasive plants, the vine didn't have to fight for its foothold across the South. It was invited in, all the way from its distant habitat in China, Japan, and the Korean Peninsula, where its rhizomes have long been crushed up as a prized source of powdered starch, known in Japan as *kuzu*. It was first formally introduced in the United States at the 1876 World's Fair in Philadelphia, where it was unsuccessfully pitched to farmers as a feed crop for cattle that was nutritious, sturdy, and easy to grow. It wasn't until 1933, when the Great Depression hollowed out the South, that kudzu would get its big break. As a clinging vine that spreads muscular root systems across vast areas, it seemed like a perfect weapon against soil erosion caused by poor agricultural practices. After an act of Congress, more than 85 million kudzu seeds were distributed to farmers across the South—a balm, said a popular radio host at the time, for a barren land "waiting for the healing touch of the miracle vine." Alabama, an agricultural and industry leader before the Great Depression, was among the most zealous converts to the new seed. But even early on there were hints. Fast-growing, thick, and turgid, kudzu, once it got stuck into the soil, was hard—impossible, even—to dislodge.

Despite a concerted campaign to cover Alabama in kudzu, not everyone bought it as an ecological cure-all. A short article in *The Birmingham News* dated May 16, 1944, detailing the state-wide practice of lining Alabama's highways with kudzu to reduce roadside erosion, reported that unnamed miscreants—"probably those who owned the property adjoining the right-of-way"—were digging up rhizomes before they could gain a foothold. "This was probably done more through ignorance than through malice," intoned the article. "They probably feared the spread of kudzu which is a needless fear." But that fear quietly spread as the full scope of what had been unleashed across the South began to reveal itself. As early as 1945, farmers started buying less kudzu, seemingly turned off by its rampant spread. By 1947, "kudzu-like" had entered the lexicon as a synonym for "unchecked." By the 1950s, masses of the vine had pooled across Birmingham, becoming the seedy backdrop to true-crime stories. Illegal gambling machines were found in a "kudzu-covered culvert" after a months-long search by detectives. Two young sisters went joyriding, only to plunge two hundred feet down nearby Red Mountain; they were saved by the soft field of kudzu the car landed on below. An ex-cop turned bootlegger was chased through a kudzu thicket by police, becoming entangled in the vines. And on it went.

By the late 1950s, the vine had become so ubiquitous in Alabama that it engulfed entire apartment complexes, erasing them in a shimmering haze of leaves and vines. *The Birmingham News,* a staunch defender of kudzu only a few years earlier, had by then switched sides and tried to rally the citizenry to action. The rampant spread of kudzu "should have sounded a shrill, high note of alarm, with calls for spades, salt, kerosene—anything that would discourage or eliminate the fearsome plant, if such exists." But the invader, everyone agreed, had already won. "Complacency at a threat such as that could well result in the discovery one of these days soon that our capital city has been completely obliterated under a blanket of green, the only movement the slight stirring of the kudzu leaves as the breezes play through Central Alabama—another Carthage, its conqueror kudzu rather than sand."

By the turn of the twenty-first century, the miracle plant had been officially reclassified under the Federal Noxious Weed Act and had earned the moniker "the vine that ate the South." *Time* magazine piled on, including the introduction of kudzu on its list of the one hundred worst ideas of the twentieth century, along with the Treaty of Versailles and the sailing of the *Titanic*. And while *Time* didn't put it this way, the mistake we made with kudzu was forgetting Albert Thellung's theory of adventive floristics. There were two kinds of plant invaders, Thellung wrote, those that require human interference to survive (which he called *Epökophyten,* combining the Greek *epi, oikos,* and *phyton* to form a word that translates as "plants upon the family's property") and those that do not (*Neophyten*: "new plants"). Epökophyten invasions can be easily controlled because the aliens rely on our continual care to keep them alive. Let the garden of your *oikos* go to seed and the Epökophyten will die. But that passive approach won't work with the Neophyten, which can prosper without humanity's helping hand. Neophyten learn quickly, exploiting gaps in their adopted habitat until they come to dominate it. Turn your back on them and the next thing you know, your *oikos,* your apartment complex, and your city are covered in a blanket of tangled green.

■　■　■

"If money isn't loosened up, this sucker could go down!" It was a rally-ing cry and a desperate plea for help, and it came from the world's most powerful person: President George W. Bush, long maligned for his malapropisms, who had in this case minced no words. The date was September 25, 2008, and a national housing crisis had boiled over on the back of trillions in bad mortgages. Bush was trying to sal-vage a seven-hundred-billion-dollar bailout package that would, he hoped, stop U.S. banks from collapsing under the weight of their own cravenly stupid investments, which threatened to drag the world econ-omy down with them.

While Bush tried to force an agreement through, Jace Goodling was desperately close to bankrupt. Since 1996, Goodling had been the proud owner of Goodling Enterprises, LLC, a Virginia-licensed Class A home builder, and he was seeing all that he had built over the past decade fall to pieces. In the wake of the housing crisis, the U.S. con-struction industry had ground to a sudden halt, leaving Goodling and other contractors readying for a hard crash.

Goodling had been doing custom home builds and by 2008 was running a healthy portfolio of jobs across the state of Virginia. "And then," Goodling says to me, recalling that chaotic time, "the recession happened. And just about overnight I went from four million in con-struction to couldn't find a screen door to fix. When that bubble burst, everything just stopped. It was like a spigot shut off."

Worse for Goodling was the fact that he wasn't alone. In fact, he had twenty-two mouths to feed. There was himself, his wife, Louise—a horseback-riding instructor at a local college, who was at risk of losing her job too—and his two devoted sons, Atlee and Clarke. And then there were the goats. In 1996, the same year he started his con-struction company, Goodling also splurged on three Kiko goats, an off-brand breed developed in New Zealand in the 1980s and first im-ported into the United States in 1992. The breed's main selling point was resilience—to parasites, disease, famine, and other daily hardships. "Which made them perfect for me," says Goodling. "I'm a lazy farmer. I don't want to be deworming goats all the time. I want the strong ones." Goodling's idea was to use the goats, which ate voraciously, to clear his property, a twenty-five-acre hobby farm in rural Afton, Virginia, that rolled up the foothills of the George Washington & Jefferson National Forests and was, when he bought it, impassably overgrown. He also figured that he might just make a little money on the side from selling breeding stock.

But then time rolled on and it was 2008, the financial crash happened, and the sucker indeed went down, with millions of Americans losing their homes, their savings, and their livelihoods. Goodling, staring down the barrel of forfeiture, was forced to take stock of his assets. There was the custom-build company. There was the bucolic farm. And there were the goats. The calculus was simple: Neither the company nor the farm yelled at him daily for feed. The goats never shut up.

One day, as he sat on the porch with Louise, smoking his pipe and sharing a bottle of wine, Goodling tried to relax but couldn't shake the same nagging questions. *How am I going to make the mortgage next month? What's going to happen next?* A proud, self-reliant type, Goodling still couldn't bear to cull the herd that he had spent so many years multiplying. Then, as he looked out at the majesty of the Shenandoah mountains that surrounded him, he thought back to how far he had come. When he arrived, the farm was a scrubby, vine-covered hillside, a dog's breakfast of multiflora rose, Chinese privet, coralberry, and every sort of fast-growing, fast-establishing tree, like wild cherry, walnut, and ailanthus—weeds, all of them, many of which were invasive species that had found their footing in the gentleness of the Virginia climate. But the goats had changed all that. "They ate all the weeds," he says, "and all the other stuff too: the coralberry, the honeysuckle, other seasonal weeds that popped up."

And then the thought arrived. "Just like a light bulb," says Goodling. "*I'm going to put the goats to work. Why not? It could be a perfect time,*" he remembers thinking, "with all the environmental awareness." Afton, where his farm was located, is just a twenty-five-minute drive to Charlottesville, home of the University of Virginia. "Being a college town," Goodling adds, "it's a ripe opportunity for more liberal, environmentally minded people." The idea of using goats to clear land would be an easy sell over there. "I figured, if one percent of the people try me, that's probably a career's worth of work right there." So, with his business failing and financial stress digging new ridges into his brow, he tried one last Hail Mary. Instead of being sold as meat, the goats would sing for their supper.

■ ■ ■

"Goats live their lives in two-hour cycles. They're up and around eating, playing, causing mischief. And then they lie down, they rest, they chew their cud, they contemplate their future mischief for two hours, then they get up. They repeat that around the clock."

"What kind of mischief do goats enjoy?" I ask Goodling.

He takes a long thoughtful pause. "Anything you don't want them to do. Left on their own, they are a force of destruction. The flower beds you were given as wedding presents years ago—they'll eat them first. Then they'll play King of the Mountain on your vehicles. They will go into any open building or space and create total chaos in there, just for shits and giggles. I can talk anybody out of goat ownership," he says. "Usually all it takes is bringing in a biblical example. Because whether you are religious or not, you know that in the Bible, God's people are always sheep, but the goat works for the other team, and there's a reason for that. Sheep will find a way to die because it's just easier that way. Goats," Goodling says, "are going to have a lot of fun making trouble before they go out of this world."

I'm trailing Goodling as he tramps up and down a long winding dirt driveway off Sugar Hollow Road in a rural corner of Virginia called Crozet. I ask him questions and he answers, walking away as his sentences drift through the early summer heat, my footsteps kicking up dust as I follow him, straining to hear. He works unhurriedly, unspooling rolls of temporary plastic fencing set through with electrified wires and placing them along the edges of the driveway, eventually enclosing a half-acre-sized thicket of vines, saplings, and tall grasses too dense to walk through. Goodling might be moving slowly, but his forty-odd Kiko goats, which are covered in various hues of brown and white fur and boast stubby horns on their long thin skulls, are quickly trampling their way through the thicket, braying happily as they nibble everything in sight. This is a typical job for Goodling and his company, which he named Goat Busters. A homeowner with a rambling rural property, a distaste for pesticides, and time to spare hires Goodling—well, hires his goats—to clear their land. Being devoured by goats is an ideal way to control kudzu infestations, given how impervious the vine is to being mowed down or torn out, and Goodling is constantly facing down thick masses of the stuff. Goodling runs electrified fencing around sections of the property within which his goats run wild. Once that section's cleared, he'll erect fencing around an adjacent section, let the goats in, and repeat, while Goodling does cleanup on the first section, pulling out rhizomes and roots, and using his chainsaw to cut invasive trees down and break up their branches, a critical piece of work that ensures the plants don't immediately spring back to life. After a time (Goodling's herd can clear three acres of thick brush in two and a half weeks), everyone walks away happy. The client's thicket is transformed into a clearing, Goodling pockets some cash, and the goats leave happy and well-fed.

Mostly. Like any business, Goodling's has seen jobs turn pear-shaped. It's just that when his go south, strange things happen. I ask him to elaborate. "Now, on any given job," he says, "I may have a couple goats that I don't take out, including my herd sire. I made that mistake once and I'll never do it again."

"Your herd sire?" I ask.

Goodling looks at me patiently. "The only goat with balls on the farm." I nod dumbly, the bleat of goats in the background cutting through the silence. "Now, if you were the one male in a herd of eighty or ninety others that are all female—with a few males without their nuts—you'd develop a pretty big attitude, walking around with those big nuts hanging down. And that big . . ." Goodling trails off, his Virginian courtesy kicking in. "And they have huge horns. Well, one time I had a pretty big brush job down below Richmond, and I decided, *Okay, what's it gonna hurt to take the buck along?* It was late in the season, and it was ready for him to start breeding."

"Right," I say.

"No, no," Goodling corrects me. "He decided that this was new territory, and he was bossing all the other goats around, and when the client came around, he went to try to get at him too." He sighs. "Well, the electric fence fortunately was doing its job, and . . ." Goodling shakes his head at the memory of a client being accidentally electrocuted. "It was just one bad experience that will never be repeated."

When he speaks on kudzu, Goodling turns professorial. "It's rooted in rhizomes that are usually four to five feet underground and about the size of a basketball, at least. And from them, there will be dozens of tendrils leading out to vines that sprout aboveground. Picture a potato, and all the eyes of a potato. Same idea," he says. "I've done many jobs where I cut a fence line around the kudzu, and if I don't cut the vines back at least a foot, when I bring the goats the next day, the vines will have already reached the fence and be grounding it out. That's how fast it goes."

"Shorting out the electricity, you mean?" I ask.

"Correct. Correct. Correct." Goodling attacks the world with a genteel confidence, a sense that anything can be bent and banged into shape with the precise amount of strength and care. That's why, in 2015, he took on a job that he might have been wise enough to refuse if he had reflected on it a while longer: ridding Birmingham, Alabama, the nexus of North America's kudzu invasion, of its most famous synanthrope.

▪ ▪ ▪

Red Mountain Park, where those two joyriding girls were saved by a bed of kudzu after a two-hundred-foot fall back in the 1950s, lies a few short miles southwest of downtown Birmingham. It's a fifteen-hundred-acre green space shaped like a long thin worm tunneling toward the heart of the city, crisscrossed with fifteen miles of forest trails. It's also a testament to the power of kudzu to not only invade landscapes but also choke them out. The park looms above the surrounding landscape, and as you approach, it almost glows green with kudzu, interspersed with stands of dogwood, Florida maple, black walnut, and Southern shagbark hickory.

But this forest on a hill is not quite what it seems. Red Mountain Park is the scene of the action for the first two of Birmingham's three-act tales of rags to riches to retreat, a place that has been hollowed out like the city itself, but in a very literal way. Jefferson County, where Birmingham is located, is the only place in North America that contains large deposits of all three of the minerals needed to make steel—iron, iron ore, and limestone—and they all lie either within Red Mountain itself or within a few miles. Prior to the 1880s, Alabama was an agrarian society with most of its economy based around cotton production. But the search for steel to feed a growing industrialized United States at the turn of the twentieth century intersected with the unexploited resources of the post-Reconstruction South. Alabama, long dismissed as a plantation backwater, a place made bankrupt by the Civil War and only recently relieved of military occupation by the U.S. War Department, was suddenly among the most lucrative mining sites in the world.

And so industry went to work. Red Mountain was stripped and more than sixty mines were dug into its earth, the iron ore extracted over and over again, until the forest was transformed into a labyrinthine network of empty space held up by underground joists and topped with a thin sprinkle of soil. At its peak, pilots used the red glow of Red Mountain's steel furnaces at night, which could be seen from miles away, as an aeronautical landmark, like an aurora borealis that never turned off. From a town of about 3,000 in 1880, Birmingham's population exploded to more than 25,000 by 1890, and reached 340,000 by 1960. It was a Southern Magic City, powered by steel to feed the nation's industrialization at home and its emergent war machine abroad.

But the magic always fades. Eventually, Birmingham's mines were shut down, the consequence of dwindling local reserves and cheaper mining in the United States' growing sphere of influence in Latin America. In the 1960s, Birmingham's role as the Southern battlefield for civil rights and the violence that followed—bombings, attacks on Black children, and the mass arrest and incarceration of civil rights protesters, including Martin Luther King Jr.—accelerated its decline from a paragon of New South prosperity to a city unmoored. By 1970, through the combined impact of urban white flight and the Second Great Migration, which saw many Black families (including Chris Schell's) leave the South, Birmingham had lost 40,000 residents within a decade. By 2023, the city's population dipped below 200,000, a 40 percent decline from its peak in 1960, and a gloomy epitaph to its faded prosperity.

The deep social and economic problems that Birmingham has faced have also made it susceptible to synanthropic invasion. On Red Mountain, kudzu quickly colonized the area around the empty mine shafts after the companies that owned them, concerned about the hollowed-out hill collapsing in on itself, planted the vine to slow the erosion of the thin layer of topsoil they had left behind. From there, kudzu crept through the park and down its steep hills, before jumping across the surrounding highways to inhabit the small homes purpose-built for miners. Across Birmingham's downtown core, the vine became a fixture in the city's tens of thousands of vacant lots.

Goodling was hired to clear Red Mountain in 2015, when one-tenth of the massive park had become impassable and kudzu, true to its reputation, was spreading rapidly in all directions. Out of desperation, Red Mountain Park staff offered him a ten-year contract to clear it before he and his goats had even set foot and hoof in Alabama. Given the size of the job, they also insisted that he maintain a herd that was at minimum four hundred strong. "And that," says Goodling with a smile, "looked very attractive." Like any small business owner, he was always on the lookout for ways to grow, and there were other benefits to a herd that size. "If I'm maintaining four to six hundred does," Goodling explains, "I'm going to be getting eight hundred to a thousand kids a year. That's a meat business in and of itself, significant enough that I don't need to move those goats to sell them—at that level, the buyers will come to me and leave me with a big fat check." It seemed like a perfect opportunity.

"But," says Goodling, "I'm a simple country boy. I wasn't ready to sign a ten-year deal with people I didn't really know, so we went for a one-year deal, on the understanding that we would try it out and see how it went." Goodling hitched a twenty-eight-foot trailer to the back of his tan-colored GMC 2500 diesel pickup, loaded eighty ornery goats into it (the other three hundred or so were gestating in their bellies), and drove nine hours down I-81 South into Alabama. He arrived in mid-October, just as the kudzu flowers, sweet and fragrant eruptions of purple and pink that grow in conical clusters until they droop under their own weight, were shedding their petals to reveal the heavy seed pods that would soon scatter the plant's genome across the willing earth. But Goodling didn't really see the flowers. All he saw was the job.

"There was plenty of food for the goats to eat," he says. The plan was to have Goodling erect fences in half-acre increments and have the goats move through the park section by section. Upon seeing the scale of the invasion up close, Goodling realized that wasn't going to work because there was no open space for him to even set his fences down. "Too much privet," he says with a sigh. "Too much kudzu." So he changed tactics. "The new plan was that the park staff would take out a forest mulcher"—a cross between a tractor and a lawnmower—"to establish our fence lines. They would send it through, we would come behind it, clean it up, erect our fence, put the goats in, and let them eat everything they could reach."

It was a good plan, and as Goodling stood his fences up, turned the electrical current on, and let his goats loose in the first enclosed sections, it seemed like it was going to work. The endless fields of kudzu, so tightly coiled together that they resembled a body of water, steadily receded with the goats' incessant nibbling, stamping, and general mischief. Soon Goodling, like a modern-day Napoleon at the head of a horned army, was taking acre after acre from the enemy, and finally bringing order to the chaos of Red Mountain.

■ ■ ■

Synanthropic plants face an uphill battle. As Albert Thellung first codified, an alien species must first get to a new environment, survive long enough to propagate offspring, and then, if it's lucky and cunning, sustain itself across a landscape. Remarkably, these stages are fractal: They neatly align with the steps a novel virus must take to get from the respiratory tract of a single bat to causing a global human pandemic. In both cases—complex multicellular organism or barely alive virus—the odds are stacked against success. The process requires so many things to go right (or wrong, depending on your perspective) that failed invasions outnumber successful ones by massive orders of magnitude. And even when a synanthropic plant can put down roots, the calculus is cruel: Not only must the environment be suitably hospitable; the invading plant also needs a tool kit of useful traits to help it secure victory.

In the case of invasive plants, two traits appear to tip the scales. In a seminal study from 1955, the English botanist Herbert Baker published the results of his year-long study of the fitness of leadworts, a thousand-member family of blue and pink flowering plant species that has managed to establish itself in habitats ranging from the Arctic to the Tropics. In what would later be dubbed "Baker's Law," the botanist discovered that the success a plant species has in dispersing itself across long distances is highly correlated with its ability to self-replicate. It makes intuitive sense: Plants that only propagate through outcrossing (the sharing of genetic material with other individuals) need a lot to go right. Upon landing in a new environment, an outcrosser needs a mate, but it also often needs insects, bats, or birds to transfer pollen so that it can reproduce. Compare that to a "selfer," which is a plant that has both male and female reproductive structures and can reproduce asexually; all it needs to spread is its own company. For that reason, a single self-propagating invasive plant can produce, on balance, 50 percent more offspring compared to a plant that needs a mate. It's an efficient shortcut, especially in that early vulnerable period after a plant species arrives in a new environment. The cost? Just like the kink-tailed mountain lions stuck south of U.S. 101 in Los Angeles, the smaller the pool of genes, the less likely it is that an organism—be it animal or plant—can avoid extinction. This means that self-fertilization, the most effective short-term survival strategy, is directly at odds with the long-term fitness of plants.

Faced with this trade-off, kudzu has one answer: yes. Though the vine primarily reproduces by cross-pollinating, it has evolved the capacity for "selfing." When colonizing a new area, a single kudzu plant will switch to a selfing mode, clone itself, and ride out the initial period of scarcity. When many plants are well established, kudzu will then switch back to an outcrossing mode and create genetically diverse offspring that have a better chance of weathering whatever nature's abject cruelty might send their way. Kudzu patches born of self-pollination are weaker, with smaller leaves, fewer fruiting bodies, and far fewer seeds (three square feet of kudzu can produce as many as 1,800 seeds or as few as 1). And while switching happens within individual patches depending on local conditions like temperature, humidity, sunlight, and the abundance of nutrients in the soil, it's also taking place at the continental level. At the northern edge of kudzu's range, which keeps pushing farther and farther into the U.S. Northeast and Canada, kudzu is cloning itself much more frequently than in the areas it dominates in the South. Despite this, across the vast North American kudzu patch, the level of the plant's genetic diversity is sky-high. In one study that sampled kudzu genomes across twenty locations in the South, for instance, botanists found that 93 percent of kudzu genes were polymorphic, meaning they were expressed differently, which is a telltale sign of an incredibly high level of genetic diversity, and far higher than many other alien invasives in the region, like cheatgrass, rough cocklebur, and velvetleaf. In the South, the last bastion of many ancient relicts, kudzu's genetic diversity and creeping dominance is further tipping the scales toward their inevitable extinction.

Worse, human efforts to stem that tide and save the remnants seem to be having the opposite effect.

Controlling kudzu by mowing down patches has only helped it become more resilient. Kudzu seeds need to be scarred to germinate, and mowing both scars and scatters its seeds, ultimately increasing the overall territory within which it can take root. Not even stripping it to its rhizomes will do much to slow it down, it turns out. In a controlled experiment, seeds were able to survive up to two months with zero direct sunlight, while removing 75 percent of a kudzu patch's leaves didn't slow its spread at all; the plant just continued creeping ever forward in all directions.

■ ■ ■

As Goodling moved his herd across Red Mountain, he was contending with a biological invader that had been invited in, had the traits necessary to weather times of scarcity, and had been given decades to spread. Still, Goodling saw no reason why he wouldn't ultimately prevail. His goats, for all their mischief, were dependable eaters, and their numbers had been rising steadily as the herd sire made his way through the does. In each new section, the herd spilled out into the brush and nibbled frantically at everything, even pulling down the kudzu lianas that had climbed trees, where the sunlight was less obscured and from where the plant could spread seed even farther than from the tendrils on the ground below. After his goats had finished their work, there was one last critical step: The park staff had to run the forest mulcher through the cleared area to grind up the potato-like rhizomes stuck firmly in the soil, which was the only part of the kudzu plant that the goats couldn't eat. It was a surefire approach that Goodling had used for over a decade, and Red Mountain had ten employees, which was more than enough to handle clearing the land after they passed through. The goats, for all their quirks, were reliable. But other humans? That's where the real chaos was found.

Unlike a lot of vocations, Goodling's can't be done remotely. To take the Red Mountain job, he had been forced to say goodbye to his wife, Louise, and their family and relocate to Birmingham, a city that he admitted he wasn't too fond of. "I'm a country boy who never wants to leave my own farm," he says. Heading down to the Deep South, he was mostly worried about race relations, "but I didn't see any open hostility." Goodling brightens. "Now, I did see a lot of stuff that I didn't expect, like people wearing guns in line at McDonald's." Being exposed to gun culture was discomfiting, but the job carried other kinds of risks. As the year progressed, Goodling's herd started to dwindle, to the point that about forty goats in all went missing—stolen, as confirmed by photos of bandits driving into Red Mountain in the middle of the night on four-wheelers, tearing up his electric fence, and making off with a few choice does. "We did have livestock guardian dogs that lived with the goats twenty-four/seven, but I have the challenge of needing to have dogs that will protect the goats but have the judgment not to be people-aggressive," he explains. "I grew up around the University of Virginia. I know what kind of stupid shit fraternity boys do. And I don't want to wake up and see my name on the morning news because one of my guard dogs mauled some UVA frat boy who was taking a dare trying to steal a goat."

While he watched helplessly as his herd got depleted, a bigger problem arose. Four months into the job, he discovered that the ten park employees weren't quite the Goat Busters support team they had claimed to be. Eight of them were full-time office staff, and another split his time between the office and the field. That left only one man in his sixties, Jeff Newman, in charge of pretty much all the outside work: maintaining fifteen hundred acres, fifteen miles of trails, dozens of trash cans that needed to be emptied every day, and bathroom maintenance across the park. "And," Goodling adds, "keeping up with the goats," which meant passing over the dozens of acres that Goodling's herd had cleared with the forest mulcher to kill the kudzu rhizomes for good so they couldn't spring back to life. Still, Newman was game, and he seemed to be keeping up okay, though Goodling's full focus was on the kudzu-covered areas ahead, not with the parts he had already cleared. By December 2015, the goats were still on track to clear the land he had been assigned, despite the setbacks. The lure of a ten-year contract was looking pretty enticing, and the meat he would be able to sell as his herd ballooned would be a windfall. But as the Yiddish proverb "You plan, God laughs" caustically reminds us, nature has a way of upending our best-laid plans.

On Christmas Day 2015, Goodling and Louise, who had joined him from Afton for the holidays, were eating a Christmas meal with some new Birminghamian friends when the weather suddenly went crazy. Everyone's attention turned to the TV, which was broadcasting play-by-play, street-by-street coverage tracking the movements of a nearby tornado—an EF2, with winds up to 135 miles per hour. "It was fascinating to this ol' country boy from the mountains of Virginia," Goodling says, before mimicking the TV weatherman. "'The tornado is moving east along Jefferson Boulevard, now at Forty-Second Street. . . . If you're on Forty-Third through Fifty-First Streets, seek underground protection! Now it's veering over toward Handley.

If you're on Handley or Sipes, run for your life! It's heading straight toward Red Mountain Park!'"

Those last three words hit him like a ton of bricks, and he watched helplessly as the radar showed the tornado passing directly over the area where he had left his goats. As soon as the weather system had dissipated, Goodling headed out to survey the damage to his herd. "I expected to find goats scattered for miles," he says. Luckily, though several sections of electrified fencing had been destroyed, the goats were too preoccupied with the new buffet of broken-off treetops that the tornado had dislodged to even notice that they were free to leave. Goodling quickly put up new fences and secured the herd. Then, he says, "We retired back to my apartment to celebrate surviving our first tornado."

It may have spared his goats, but the Christmas Day tornado caused widespread flooding, power outages, and the destruction of seventy-two buildings in Birmingham. Across the fifteen acres of Red Mountain Park, trees were downed all over, leaving park staff to scramble to clear debris from its crisscrossing trails. After such a thorough trashing, Newman, the only park staff assigned to help Goodling and his herd, was on full-time tornado cleanup. That was a problem: Goodling couldn't set up new fence lines without Newman running the mulcher to clear a path, which meant that he couldn't move his herd forward. But when you have 416 goats to feed, standing still in a field that has been stripped bare of every leaf and vine isn't an option. The pressure—braying, trampling, butting—for fresh forage is overwhelming. The only logical solution, Goodling figured, was to see if any of the previous sections that his herd had passed over, which still had cleared fence lines, could use touch-ups. So he went back to the beginning to see what had become of his work.

Goodling expected to find some growth, but the scale of the disaster was heartbreaking. Instead of scrubby grass and short saplings liberated from the kudzu's grip, it was as if Goodling had traveled back in time to before his arrival in Birmingham. "They never got around to coming back and forest mulching the areas that we had cleared. So, what happened?" Goodling asks rhetorically. "It grows back. It's just a nasty-looking mess." The landscape was a tidal wave of vines and leaves that transformed branches, boulders, and the exposed roots of dogwood trees into a blur of bottle green. Worse, as Goodling approached, he noticed that the kudzu had grown back even thicker than before, a result of the seeds that were scattered by the goats while they gorged on the vines. In that moment, it was with some bitterness that he realized how deeply futile his efforts had been. By leaving the rhizomes intact, it didn't matter how many acres of kudzu he and his goats mowed down. The plant was always going to win.

After the tornado, Goodling spent the next few months re-clearing land, all of which had been covered over with thick masses of kudzu. By May 2016, eight months after he had moved down to Birmingham, he and his goats were going back to areas they had previously cleared for the third time. It was as if Goodling was battling a tumor using a blunt and ineffective chemotherapy. No matter how hard he tried to strip the land, there was no way to dislodge the growth; all he could do was slow it down for a moment before it roared back to life. Contraction, clearing, renewal, aggression—if you've ever lost someone to an unrelenting malignancy, then you know the roller coaster of success and failure. No wonder the *Birmingham Post-Herald* once dubbed kudzu "the vegetable form of cancer."

By August 2016, Goodling was heading back up I-81 toward the family farm in Afton, Virginia, a defeated man. The combination of kudzu's relentless mobility and a poorly equipped park staff had left Red Mountain in worse shape than when he had arrived ten months prior. As he raced home, pulling hundreds of goats packed like sardines in the trailer behind him, the highway ditches that lined his path were covered in kudzu, as if the plant were sending him away, a conquered vassal, with a final humiliation.

■　■　■

More than a decade later, kudzu still covers Red Mountain Park. Jeff Newman, the sole park employee tasked with clearing it, does what he can. But it'll be his last year, he confides in me; he's turning seventy soon, and the physical punishment is just too much. I follow him up to the top of a hill through a narrow path that he has cleared. We are surrounded by a green smear of kudzu with no discernible features, not even a sapling sticking out from the entanglement. I express shock at how unrelenting the plant can be. Newman points south, and I see that the kudzu continues on the hillside for miles on end: the same dark greenery, its intensity drawing you in like a black hole.

The work is constant. "You can cut it back," Newman says as we tour the ruins of an old hoisting station that was used to lower iron ore from the top of the hill during Birmingham's steel-making glory days, "but that won't stop it." Newman recently used a riding mower to set a precise line around the ruins, leaving a gap of a few feet where scrubby grass has popped up between the old concrete husk and the mass of kudzu that surrounds it. You can practically feel the vines crawling across the open space. "Sometimes I'll come up here and pour Roundup"—a potent herbicide that has spawned tens of billions in lawsuits because of its link to a range of cancers—"and gasoline on the root crowns. It's the only way to kill it. At least for a while."

From the hoisting station ruins, which are littered with small, smooth filaments of shattered glass ("in the 1970s, the mining company came and smashed it up rather than paying taxes"), Newman leads me over the hill to a sprawling open space covered in a thick and impenetrable layer of kudzu. "Jace cleared the entire area we're on," Newman says. "It was down to the dirt. And then it rose from the dead by the next summer."

During my time in Birmingham and crisscrossing Alabama's highways, kudzu is a constant reminder. It thrives in every roadside ditch, in many of the city's twenty thousand vacant lots, and on the stubby hills behind every parking structure. And as I wander the edges of Red Mountain Park with Newman, it's everywhere here too. Until it isn't.

On the edges of the park, the kudzu is like Velcro, its vines interlocking, covering everything. But after a few hours of following Newman along the park's circuitous trails, we finally weave our way to the center, where stands of dogwood, hickory, walnut, and maple grow. These are young trees, all of them, thriving in the aftermath of the apocalypse the miners wrought on the mountain. Their leaves dapple the sunlight, a welcome canopy, so contained and calming compared to the barely there urban canopy that Birmingham, with its spaced-out downtown design, affords. On the edge of the trails, the kudzu mobs the trees, sending lianas up their trunks and branches. It's a worrying sight. But as we step off the trail and head deeper into the forest, a strange thing happens. Once the canopy's shade solidifies, about ten feet from the path, the kudzu begins to recede from sight. Twenty feet in, as we pick our way over crumbling logs and a mess of other invasives, including privet, mimosa trees, and imperial bamboo, the kudzu is no more, as if the forest itself has the power to melt the vine away. It's an astounding trick, and wholly unexpected.

For all its success as a synanthropic invader, kudzu's apparent ubiquity might just be a side effect of our urban myopia. In cities and the open arteries that connect them, kudzu appears unstoppable, bathed in direct sunlight and exploiting disturbed and emptied urban landscapes. As soon as a canopy closes above its leaves, the plant is halted in its tracks, made as harmless as a newborn kid. Kudzu is heliophilous, its small, thin leaves designed to maximize its rapid growth under direct sunlight. In full shade or even low light, though, kudzu rapidly reaches its "photosynthetic compensation point," meaning the energy it uses to photosynthesize becomes greater than the benefits of doing so, which ceases its growth. In the deep, canopied shade of forests, its massive rhizomes also become a drag on its survival, and they end up shrinking faster than the leaves themselves, which causes the plant to die out. Most of us rarely find ourselves penetrating that far into forests. Instead, our idea of what plants can and cannot do is shaped largely by what we see with our own eyes in the early successional habitats we call home. In Birmingham and other cities in the South, what people see is a lot of kudzu.

This myopia is borne out at the macro level, where the scale of kudzu invasions has been drastically downgraded in recent years. In 2004, kudzu was estimated to cover 7 million acres and be spreading at a rate of almost 124,000 acres per year. But a 2010 estimate by the U.S. Forest Service found that kudzu covered only about 91,000 hectares of forestland in the United States, or roughly 0.1 percent, and that its spread across the continent was about 2,500 acres per year, not 124,000. That doesn't mean kudzu isn't an ecological threat or a supremely well-designed invader. While it isn't well suited to penetrating deep forests where canopies are thickest, it can still poke away at their edges, strangling saplings, smothering delicate relicts, and causing woodland borders to retreat, albeit more slowly than previously thought. All in all, though, it appears the vine is much better equipped for urban warfare than guerrilla tactics in the wild. This is cause for some cautious optimism on two fronts. First, the danger that kudzu poses to the natural world appears less total than most of us had assumed, making it primarily an urban nuisance rather than an unstoppable destructive force that will inevitably drive native and endangered plant species to extinction. Instead, with the right strategies, we may just be able to stem its spread. Second, this battle will inevitably take place in cities, where kudzu has found sanctuary under the direct sunlight spilling through the urban canopy. It's a hopeful reversal: Rather than resign ourselves to the loss of forests from kudzu's voracious spread, populating our cities with green and forested spaces might just end the nightmare of the noxious weed. Along with a steady stream of goats, of course.

▪ ▪ ▪

Back at Goodling's acreage, kids tumble through grassy slopes, their stubby horns poking through glossy white fur. He guides me to the top of the hillside farm, past a dilapidated wrought-iron-enclosed cemetery with centuries-old gravestones, and we stand at the edge of George Washington & Jefferson National Forests. We turn to look out across his land and the sprawling vista beyond. It's an incredible view. The afternoon sun is shining matte-like, casting warm viscous light on Virginia's rolling hills, which descend into the distance evermore. Goodling is content. There were no mishaps at the worksite at Crozet, where the bulk of the herd is still munching away, and the client seems happy enough so far. Still, Goodling knows that no matter how large or hungry his herd is, the kudzu and other neophytes like it will always find a way to root, establish, and spread. And while nibbling at the margins of a blight might be demoralizing for most, Goodling has evidently attained some kind of peace with the job. His farm was impassible when he found it, but it's a pristine jewel now, the war with invasives limited to its edges. I watch him as he stares out at the hills, his long, lean frame bent slightly as he takes it all in. I'm reminded of Sisyphus, the cunning Greek king condemned for eternity to roll a boulder up a hill only to watch it roll down again, his punishment for believing himself capable of cheating the forces of nature. "The struggle itself toward the heights is enough to fill a man's heart," wrote Albert Camus in *The Myth of Sisyphus,* trying to make sense of why humans keep living in the face of existential meaninglessness. "One must imagine Sisyphus happy." Goodling, a patriarch who revels in a joke, is a good fit for Sisyphus, though his lopsided battle to clear kudzu is, in his mind, less a punishment than an opportunity. Later, I ask him whether he's ever read Camus's book. "I have not," he replies. "However, knowing what a fan Camus was of the absurd, I could believe he was somehow inspired by goats."

Chapter 6

The New Animism

All the forces of nature are going their own way.

—John Burroughs, "God and Nature"

Animal Invasions Create Chaos and New Kinds of Order.
As a general rule, synanthropic invasions only succeed when the
invading organisms are benign enough that the harm they do us is self-
limiting. Once it spills out of control, humans are forced to counteract
and drive the invader out. Rarely, though, synanthropes will descend
upon cities in high numbers, with unchecked aggression and with
sufficient intelligence to outwit our efforts to turn them back. Among
highly intelligent social animals, these invasions can bring about
episodes of violence driven by dramatic behavior changes. And once
a new behavior radiates out through social networks, synanthropes
can even reorder the hierarchy between species. When enough of these
behavioral adaptations accrue, invasions can herald the beginning
of new social orders, cultures, and animal societies.

■ ■ ■

In 2020, a ship captain watched in amazement as his vessel—a small fishing trawler traveling through the Strait of Gibraltar off the coast of Spain—was encircled and rammed by a pod of orcas. It was a strange and disturbing encounter, and proof of nature's spontaneity. Half a decade later, what seemed like a one-off event now appears to be something else entirely: an evolutionary adaptation heralding a time of increased conflict between humanity and the natural world. Since that first attack, orca vandalism across the Iberian Peninsula has become more efficient, with the whales coordinating to disable boats in ten to fifteen minutes instead of the hour or more it took when the attacks first began. They are also more common. During the summer months, when orcas and fishing vessels vie to catch the region's bluefish tuna, attacks happen almost daily.

Orcas are large animals, the largest in the dolphin family, each about the size of a city bus, and they are no stranger to humans. In Australia, the Thaua people, an Indigenous nation that inhabits the southern area of what is now New South Wales, long hunted baleen whales collaboratively with orcas, the latter of which would herd the baleen into shallow waters to allow Thaua hunters to spear them. The Thaua would get the meat, and the orcas would get the tongue. After colonists arrived in Australia, commercial whalers operating in the region continued the compact. Orcas would guide their vessels to baleen whales, the whalers would kill them, and the orcas would get their delicacy. It became known as the "Law of the Tongue."

This mutualism is nowhere to be found among orcas in the Strait of Gibraltar, which have carried out more than seven hundred attacks since 2020, some of which have ended with some vessels disabled or even sunk to the bottom of the ocean. It's a startling new dimension in our long relationship with the marine mammals. It's also a puzzling one: Why would orcas suddenly target humans? Unlike the giant Pacific octopus in Seattle, orcas aren't a primarily synanthropic marine creature, but in this region—one of the most active fishing areas in the world—they are constantly navigating waters rife with human activity.

In fact, the Gibraltarian orcas have become experts at plundering tuna from the many fishing trawlers that work the area, suggesting they receive some benefit from living near our species. If this were about food, the orca campaign of aggression would exclusively target fishing trawlers, but it doesn't, with the cetaceans overwhelmingly targeting small sailboats instead.

Some have called it revenge for the damage that humans have done to the habitat and the orcas themselves. It is a grim list: Orcas are frequently wounded (sometimes fatally) by underwater boat propellers; pollution has caused chemicals to leach into their bodies, which, as Eliza Heery noted, can cause stillbirths; they are sometimes shot for interfering with fishing operations; noise pollution dampens their capacity to communicate via echolocation; and they can be drowned after becoming entangled in commercial fishing gear. Marine biologists who study them deny that cetaceans could be as vindictive as humans, though the notion of revenge is not as crazy as it might seem. The first Iberian boat attack was carried out by a female orca who had head wounds consistent with being lacerated by a propeller, and it was she who then taught her pod how to disable, dismantle, and sink vessels.

Orcas are deeply social animals, known for their expertise in pack hunting, which requires a high degree of complex communication and rapid learning. They're one of the few animals to prey on whale sharks—the world's largest shark at sixty feet long—and do so by ramming them at high speed before rapidly working together to flip them onto their backs, which induces "tonic immobility," a state of pseudoparalysis that stops them from diving. Once immobilized, orcas have been observed tearing open the belly of the whale shark, gorging themselves on the liver and the fatty nutrients it contains, and leaving the rest of the hulking carcass untouched. From my terrestrial vantage point, it feels like the oceanic equivalent of a mic drop— showing off for the thrill of it. Marine biologists have a different take. Spending time and energy chewing through the tougher and more sinewy skin found elsewhere on the whale shark might not be worth the trouble for orcas, who can live into their sixties but are born with only one set of teeth.

Yet even the most experienced marine biologists will admit that deciphering the behavior of orcas is often a fool's game. In the summer of 1987, Southern Resident orcas, a closed society of three matriarchal orca pods living in the Salish Sea off the coast of Vancouver, Canada, spontaneously began to wear dead salmon as hats. Yes—that is not a typo. The behavior began with one female orca from K pod (the smallest of the three Southern Resident groups) but quickly spread to other pod members. Within five weeks, orcas across the entire Southern Resident society had taken up the practice. Biologists called it a fad—after all, it ended abruptly before the summer was even through—but could offer few concrete reasons why the animals would wear their food on their heads. Perhaps, they speculated, it was a display of high food availability, though 1987 wasn't the only year when salmon were abundant in the region. This is the same group of orcas, though, that also occasionally wrap themselves in fishing lines on purpose, relocate crab and prawn traps for the fun of it, and play with porpoises until their newfound toys die of exhaustion. With orca societies, behaviors come and go, often falling out of fashion after a few weeks. In the specific case of the salmon hats, though, we may have witnessed the first retro fashion statement ever recorded in nature, decades after it originated. In 2024, thirty-seven years after the fad began and ended, Southern Resident orcas spontaneously began to wear salmon hats once again. Whether it was ironic or not, we'll never know.

Cetaceans have long stood out for their complex cultures, and orcas are no exception. Whale songs, linguistic marvels that include regional dialects that evolve over time, have come to exemplify their intelligence as well as the deep role social learning plays within their societies. The sinking of vessels off the Iberian Peninsula is further proof that learned behaviors, even violent ones, can quickly become cultural movements among big-brained mammals, so deeply embedded that they are impossible to stop. That's a terrifying prospect for those that must share space with them at sea. But what happens when animal societies evolve within cities?

▪ ▪ ▪

As I drive toward al-Baha, I reflect on the warning that a local gave me as we exited the plane and walked along the barren airstrip toward the small single-gate terminal. "If you give dogs food, they become your friends," he said with a wry smile. "Sure, give a monkey food. But he'll never be a friend."

At first blush, the mountainous landscape I'm driving through exudes the same lonely desolation as the silty floor of the Salish Sea. It's barren, the rocks gravel gray or a matte salmon pastel. Dotted along the rugged hills are darker oval boulders that punctuate the gray scale, looking as if they were dipped in water before being tossed unceremoniously across the land. This is mountain scrub, an open countryside with little respite from the elements and only the sparsest vegetation, consisting of thin-limbed acacia trees, their leaves drying out in the sun, and parched grasses the color of straw.

The sprawl emerges miles from the city center. The desert hills stretch endlessly here, so space is no constraint. New builds set apart from one another at seemingly random distances begin to break the monotony of the horizon. It's impossible to tell whether they are being completed or have been abandoned and whether those squat concrete walls, some with domed windows and latticework, have been sitting idle for days, years, or decades. New or not, the hastily constructed edifices accumulate as I drive, the telltale colors of concrete, rebar, and sunlight on reflected glass tracing my peripheral vision. These are the raw materials of early successional environments frozen in time that mark the presence of human communities the world over. Besides the unfinished cubes, a few minarets are the only other feature to pierce the sky. And just as I enter the city, a wall of clouds looms, not high but dragging itself across the ground. For a moment the sun is covered over, the world becomes the texture of TV static, and the sepia-tinged landscape becomes entirely monochromatic.

It's a place that feels entirely off the grid: a lonely city set amid mountains, its buildings a polite distance from one another, as unplanned, it would seem, as the path of fallen leaves. And it's a quiet place, intentionally so. Beyond the many restrictions in Saudi Arabia (no alcohol or drugs, no displays of affection, no political criticism, and most women covered from head to toe in the flowing folds of an *abaya*), al-Baha is a small city in the smallest of the kingdom's thirteen provinces. It's a popular tourist spot among Saudis, one of the few places where you can walk through forests, wade into waterfalls, and experience elevation (at 2,100 meters, al-Baha is roughly as high as Mexico City), and it's a short drive from the dappled turquoise waters of the Red Sea. With a population of about ninety thousand, al-Baha is set within the Sarawat Mountains, a range that runs north from Mecca, that holiest of cities, all the way to Sana'a, the capital of Yemen. The design of the city, adapted to the sheer cliff faces here, is as vertical as it is horizontal, and few of its roads run straight, instead curving themselves up the mountainous edges like lianas on tree trunks.

It's steep here, the negative space as striking as the mountains themselves, their slopes so straight and forbidding that it feels as if they are about to slice open the sky. While the cars move fast along the roads, life moves slowly. But the general feeling of lassitude is an illusion. Because here, in the Sarawat Mountains, the organized chaos of nature risks upending not only the lives of the citizens of al-Baha but also the most holy tenets set down by the Prophet himself.

▪ ▪ ▪

It is midday, and female students are in a central courtyard of the Al-Baha University, eating lunch, gossiping, and lounging on the well-manicured lawn. In broad daylight, wholly unafraid, the attackers come over the walls, rapidly moving brown blurs scaling expertly down in waves. There are a few dozen at least—large hairy beasts moving with purpose, their silence contrasting with the feral cries they provoke from the young women. All hell breaks loose. Beyond the screams escaping from hundreds of mouths, are the gasps, the choked-down tears, and the quiet whimpering of animals—humans—that know they are trapped and at the mercy of predators more powerful, ruthless, and cunning than them. The baboons, teeth bared, gallop across the campus, ruthlessly searching for food. Luckier students flee indoors to classrooms, but others are too slow, and the monkeys overtake them, scratching at their bags and tearing them open, sometimes slicing their skin. After the initial barrage, the situation falls into a tense and terrifying rhythm: The students cower, watching, while the baboons patrol the courtyard with utter confidence and single-minded ruthlessness. Both sides know the game by now. The baboons have won the day, overwhelming the school without any resistance. All that the humans can do is wait as the attackers turn over garbage cans, empty out backpacks, and take their turn lounging on the well-manicured lawn while grooming one another serenely.

There's something familiar here, a deep biological memory of what it is to be caught in the dance between predator and prey. There is premeditation, surprise, and the terror at seeing cold-eyed creatures betraying no empathy while projecting an unshakable superiority. Our rarefied evolution was meant to bring our species beyond moments like these, when nature reaches out and grabs us with impunity. But that's not where the people of al-Baha find themselves. Instead, the number of baboon attacks on schools in this region of Saudi Arabia has skyrocketed over the past two decades, leaving the autocratic kingdom grasping for answers. From Mecca in the north all the way to Abha in the south—a distance of roughly six hundred kilometers—the towns of the Sarawat Mountains are captive to monkeys that have spurned human rules and instead turned to violence to take what they want.

They're not alone. Like all urban spaces, al-Baha is home to multiple species. As the population of urban baboons has exploded here, the primates have discovered that humans are not their only antagonists. Across this mountainous city, resident packs of feral dogs (*Canis familiaris,* "family dog") roam alleyways, public parks, and highway shoulders, gorging on human food waste. Unlike its human population, al-Baha's feral dogs are unafraid to meet primate incursions with the desperate violence of an unwanted scavenger that must fight to survive every day. This sleepy city is the backdrop to an interspecies conflict destined, it seems, to escalate far beyond human control.

■　■　■

Dr. Ghanem al-Ghamdi greets me warmly in the lobby of Cloud City Hotel in al-Baha, raising his hand in a meta-communicative signal. He is adorned in traditional Saudi dress: a *thawb,* a long, loose-fitting white robe that covers his arms and falls past his knees, and a red-and-white-patterned keffiyeh draped over his short-cropped white hair, the head covering held in place by two thick black fabric rings. As we make our way outside to his jeep, al-Ghamdi is quick to point out that the attacks here have deep roots. "Sixty years ago," he says in a rich and gravelly baritone, "farmers were already complaining about baboons." In the 1960s, al-Baha's economy was predominantly based on agriculture, owing to the rich soil and relatively mild and rainy conditions afforded by the mountains, which made it a breadbasket for the Saudi state and its nearby religious capitals. "Al-Baha used to be rich in cereals and sent them to Mecca," al-Ghamdi explains as he settles into the driver's seat. "Now we send Starbucks: café lattes and Americanos," he adds with a laugh. In the 1960s, al-Baha was barely a town; the region was sparsely populated with farmers, many of whom had only recently established permanent outposts. Al-Ghamdi's family was among them. His grandparents were nomadic sheep farmers, and his father, who was a teacher but maintained a hobby farm, was the first of his family to settle in one place.

The history of al-Baha's human population closely mirrors that of its other primate: *Papio hamadryas,* better known as the hamadryas baboon. One of the world's six baboon species, the hamadryas (the name of a tree nymph in Greek mythology) is indigenous to the Sarawat Mountains, as well as to Ethiopia, Eritrea, and Egypt, where it is locally extinct. The ancient Egyptians, dating back to 3150 B.C.E., worshipped hamadryas baboons in the form of Babi, a bloodthirsty and sexually ravenous monkey god who devoured the souls of the sinful after their deaths. At other times, the baboons represented Thoth, the god of wisdom, because of their sophisticated social life and penchant for congregating in groups to chatter. Like all other baboons, the hamadryas originated in Africa; unlike all the others, the species left the continent, finding its way to Arabia between 150,000 and 31,000 years ago via a temporary land bridge at the southern tip of the Red Sea, which emerged during a period of global sea level decline. (This bridge, which ran between modern-day Djibouti and Yemen, is thought to have carried early *Homo sapiens* out of Africa around the same time.) After leaving the continent, hamadryas baboons found an ideal habitat within the Sarawat Mountains, including cliffs that provided areas for sleeping and protection from the Arabian leopard (fewer than two hundred of which now survive), the striped hyena, and the Arabian wolf, and with enough rainfall to keep them alive.

For tens of thousands of years, hamadryas baboons have occupied this thin band of mountainous territory, about six hundred kilometers long and less than one hundred and fifty kilometers wide, in complex and fluid social structures. The only baboon species to be entirely patriarchal, hamadryas organize into "one-male units," which typically include up to nine females and their offspring. Frequently, though, these units merge into clans of a few dozen animals, which provide greater protection for small-scale hunting or scavenging forays. Clans sometimes assemble into bands of as many as thirteen one-male units, which gives the monkeys an overwhelming advantage when raiding large-scale food supplies. In times of abundance or peril (and for al-Baha's baboons, it has been mostly abundance), clans form a troop containing hundreds of baboons, a mob that acts by its own logic and erupts across the countryside with overwhelming force.

A typical day starts on a cliff at sunrise, where, in the absence of the tall trees of the African savannah, a troop of hamadryas baboons will sleep together for protection and warmth. As day breaks, the animals leave their cliffs and fracture into one-male units that play, copulate, and groom. A few hours later, the troop leaves the sleeping site together, and bands will set off in different directions at specific intervals to search for food. As the day progresses, bands splinter further into clans, and further again into one-male units, depending on the relative scarcity of food and water. As the sun begins to drag itself down toward the sharp peaks of the Sarawat Mountains, the units return to the same sleeping cliffs, where they groom and play and copulate again, before materializing into a troop and falling asleep under the unblinking stars.

Hamadryas baboons in Saudi Arabia travel as far as thirteen kilometers per day on routes that would have surely intersected with those of nomadic sheep farmers like al-Ghamdi's grandparents, who crisscrossed the Sarawat Mountains for thousands of years. And while conflict between the roving troops and the nomads occasionally occurred, neither side would have had much use for the other. But all that began to change on March 3, 1938.

On that day, Saudi Arabia struck oil at Dammam No. 7, a well on the edge of the Persian Gulf. It would mark the beginning of the reclusive kingdom's emergence as a global authoritarian power and one of the richest countries on Earth after its formation only six years prior following a brutal thirty-year conquest by its king, Ibn Saud. At the time, fewer than 3 million people were scattered across Saudi

Arabia's vast, mostly desert, territory. Steadily, though, as the war faded from memory and oil wealth lubricated Saudi society, towns expanded, cities were built, and a largely nomadic culture (in 1950, less than 10 percent of Saudi people lived in urban centers) was transformed into a modern nation state overseen by an iron fist.

Even a backwater like al-Baha wasn't immune to the change that oil wealth produced. Ninety-five percent of Saudi Arabia is desert, but the fertile, mountainous soil and humidity in al-Baha turned out to be perfect for growing grains, vegetables, and fruits, including olives, grapes, pomegranates, figs, apricots, peaches, mangoes, guavas, almonds, apples, and bananas. Rising urbanization and a surging population incentivized the region's nomads to settle into sedentary farming plots, transforming this scrubby landscape into a fecund paradise. What they didn't anticipate, though, was that another nomad would soon find it liked sedentary living as well. Shortly after farms were established in the 1960s, hamadryas baboons discovered that massive quantities of food were reliably available in the same place, on a set schedule, and with little in the way of protection. They began to make incursions. The solution to the baboon problem back then was simple. If a baboon approached you, you shouted at it or threw a rock. If a baboon raided a farm, it was shot and killed. That strategy made sense when human settlements made up only a small fraction of the rocky landscape here. But as urban sprawl pushed towns and cities out into the hills, plains, and valleys that make up the hamadryas baboon's home range, the primates were forced to choose: flee farther into the mountains beyond the narrow habitable range or adapt to the organization that humans had imposed on the land.

■　■　■

Al-Ghamdi is the acting president of Al-Baha University, a role that he's evidently still getting used to. None of this was ever part of the plan. He trained as a veterinarian and was originally lured into the field owing to his love of camels. "Baboons," he admits, "were not on my list at all," but he was called to the work given the scale and seeming intractability of al-Baha's human–baboon conflicts. We've been navigating side roads, but now al-Ghamdi veers onto the city's main highway, the jeep planted firmly in the middle of two lanes of traffic as we wind our way up through the thin and hazy mountain air. The region is less dependent on farming now, but the threat from baboons has only accelerated. And as someone who works at a university, al-Ghamdi has been a target of a school attack himself. "They attack student food bags," he explains, "and the café, anything like that. Then they go to an open area searching for waste before going into the hallways and the classrooms. And while they're in there on that trip, they attack kids and students, or even us, the residents of the university campus." As he drives, I see al-Baha blow by, my eye drawn to familiar totems: Baskin-Robbins and Starbucks, interspersed with signs in Arabic and one in English for a fast-food joint called Glitch Burger. "Sometimes . . . How would I put this? We are in the conflict zone with them. Then, people get into trouble, they have physical injuries; some people have been hospitalized after attacks." Despite their cruelty, al-Ghamdi doesn't fully blame the baboons for the violence. "You need to understand their behavior," he says. "They are trying to feed their babies."

As we drive, I ask al-Ghamdi whether there was anything that prompted the school attacks, which started in the late 2000s and ramped up significantly in the ensuing decade. "Well," he says, "before the baboons, we had this same issue with stray dogs." It's impossible not to notice the dogs roaming the streets here. They're large, about the size of Labrador retrievers, most of them with blond or black fur and muscular builds, and very unlike the sick and malnourished canines I've seen in other parts of the world. While they travel in small packs made up of two or three adult dogs along with juveniles, they also run together in greater numbers at times, especially in the most nutrient-rich parts of the urban habitat, where they violently protect their access to sources of food waste from baboons, often successfully. When bands of primates hundreds deep descend on a contested hot spot, the dogs are forced to retreat. In other cases, the baboons have been helped by an unwitting accomplice.

The Al-Baha University campus, a large flat green space on the edge of the city, has long been a favorite feeding site for stray dogs, where they scavenge off the considerable litter left by students. Or at least they did. Aggressive and prone to rabies infection, the dogs were a nuisance to students and a blight on the reputation of the university, which hosts specializations in animal husbandry and wildlife management. For a university training the next generation of animal experts, failing to quash a local invasion has deep reputational impacts, so al-Ghamdi and other faculty members were tasked with removing the stray dogs from the campus. This involved putting up fencing, hiring security to patrol the perimeter, relocating or euthanizing dogs that refused to leave, and creating a culture of hypervigilance to immediately respond when dogs were found onsite. The plan succeeded spectacularly. But no one saw the greater game being played. "When we solved the issue with the dogs," explains al-Ghamdi, "that's when the baboons started to come to the university."

It turned out, he says, that the dogs had been acting as sentinels against wider-scale incursions by hamadryas troops. Aggressively territorial, the dogs had been holding baboon incursions at bay, their ferocity giving the primates pause. By removing the dogs, the authorities inadvertently gave the baboons the keys to the gates, establishing the conditions for baboons to mount violent attacks on its students. It's a twist on the human shield phenomenon that Chris Schell described; here, though, humans play the part of the pigeons, oblivious to the fact that feral dogs were protecting them from an even greater menace. What had seemed like a successful mission to reclaim human sovereignty had only strengthened nature's hand.

School campus attacks, while frequent and terrifying, are only one dimension of human–baboon conflict. In al-Baha, troops have taken to sleeping on the cliffs directly below the city center, emerging every morning to take their pick of the food waste stuffed into garbage cans across town. Like stray dogs, the primates are viciously territorial, and food hot spots like supermarkets directly intersect with the places humans are most likely to frequent. That's led to seemingly unprovoked conflict. In one incident in 2022, a man was brutally attacked as he exited his home, which happened to be located near a local supermarket. Mere steps from his house, he was knocked to the ground by dozens of the animals, which clawed and bit his face and hands (their sharpened canines can grow two inches long and they have a bite force three times that of humans). As he lay screaming, neighbors rushed to fend off the troop with sticks and improvised weapons, successfully driving them away. After two days of treatment in hospital, the man returned home, only to find the baboons outside his home once again. They had, it appears, been waiting for him.

Complicating matters, al-Ghamdi explains, are long-standing Islamic beliefs rooted in the concept of *sadaqah,* or a duty to perform charitable acts, which the Koran extends to all living creatures. "Some people," says al-Ghamdi, "think that they get rewards from God for feeding animals, but this is a poor misunderstanding of the situation." That's led people here to feed baboons, especially the frequent Saudi tourists, which has turned the animals into a local attraction. In Taif, a nearby tourist hot spot overrun by the primates, locals were dismayed when the tourism commission used a baboon as its emblem, essentially encouraging people to interact with them. To make matters worse, the Taif government even built a turnoff near a popular fruit market so that people could park their cars and feed baboons out their windows. It was a baboon drive-in, with daily scenes of children passing bananas and other fruits through rolled-down windows to the stone-faced monkeys, who ate them on the car hoods. It is odd, to put it mildly, to see five-year-old children happily hand-feeding baboons, while mentally reviewing footage of the animals cornering screaming teenagers in schoolyards before drawing blood as they claw at their bags.

Fifteen minutes into our drive and we have entered Raghadan Forest Park, a manicured recreational area that sits at the vertical apex of al-Baha's city center and is replete with well-pruned acacia trees, a zip line, a mosque, and a petting zoo. It's popular among local residents and a favorite of its second-most-populous primate too. We pass a small pack of feral dogs on our way in, and al-Ghamdi offers a subdued wave to a team of maintenance staff before driving slowly through a parking lot lined with a small orchard of ornamental trees. There, we come across a clan—two one-male units, about thirty animals—roaming around for food. "Look at this big boy in the dining room," al-Ghamdi says humorlessly, pointing to a large male, shaggy and severe, sitting in an overturned garbage bin with plastic waste strewn all around him, while two females pick at the remains. I look farther up the walkway; every twenty paces or so lies another overturned blue bin, as far as the eye can see. Evidently, the clan has been here for some time, making its way methodically through the park's bounty, feasting on the best of what urban living can afford.

Al-Ghamdi explains that this site is only one of many baboon hot spots in town. Until recently, though, nobody knew quite how many hot spots there were, nor how many baboons roamed al-Baha. The last reliable census was done in 1992, and it found fewer than three hundred baboons across the town and its satellite villages. In 2021, amid the backdrop of a rising tide of primate violence, al-Ghamdi set out to update the record. And though he was aware that there were hundreds of the animals, even he was surprised by what he found. In Raghadan park alone, al-Ghamdi's survey team counted a resident troop of roughly six hundred baboons—more than twice as many in just one park than the entire regional baboon population two decades earlier. The troop had grown in size and girth through a daily routine of sidling down from Raghadan park to gorge on garbage outside a nearby mall.

One image his team captured shows a man tossing a full-to-bursting black garbage bag into a dumpster behind the mall's major supermarket; surrounding him are more than twenty baboons packed together tightly, some balanced on the dumpster's ledge, others crawling onto a shopping cart, all watching the arc of the airborne waste with quiet intensity. A single male, his dusty gray mane, red rump, and large muscular build setting him apart from the others, is positioned in the prime spot on the lip of the bin, ready to rend the plastic bag apart and feed on the waste.

All in all, al-Ghamdi and his team discovered more than one hundred baboon hot spots, many of which harbored hundreds of the animals. That placed the total number at ten thousand or more in and around al-Baha, equivalent to a three-thousand-percent increase in their numbers since the 1990s. To predict the areas in which human–baboon conflict would most likely arise, al-Ghamdi's team then mapped the locations of the hot spots and drew buffer areas around them in red that corresponded to the distance the animals were known to travel each day. Much of the town of al-Baha was covered in overlapping red circles, meaning baboons were regularly traveling to many of the areas that the town's humans most frequented. But when he extended the buffers across the maximum hamadryas home range—thirteen kilometers—the entire region became awash in red, with every nearby village and town subsumed within their daily wanderings. When al-Ghamdi reduced the buffer size to roughly one and a half kilometers, on account that urban baboons might travel far less than rural ones, the results were still shocking: The entirety of al-Baha proper was covered. The city had become a baboon paradise.

It's not just that the numbers of hamadryas baboons are rising (though they are, two and a half times faster than the human population here). Since the 1990s, al-Ghamdi and other observers have also seen baboon behavior change dramatically. Until then, baboon incursions into human spaces mostly took the form of raids on farms and other isolated outposts, with the animals rapidly striking, eating as much as they could, and fleeing quickly if they faced reprisals. But over the past two decades, ample food sources in cities have led to the emergence of a new baboon culture that has made them even harder to contain. Instead of sleeping on high cliffs in uninhabited mountain outcroppings a few kilometers away, baboons now sleep on cliffs directly below the downtown core—so close that it takes only a minute or two to enter the city. In the mornings, the animals first go to commercial areas and parks to clean out the trash cans. Then, explains al-Ghamdi, they move on to homes, most of which are town houses with varying levels of security. "So, even though they have high walls," he says, "and even though some of those walls have electric wire to keep them away, the animals still manage to jump over them or pass through if there's a small opening in the fence. Then they'll enter the kitchens and eat everything." In the late afternoon, the baboons descend on grocery stores and restaurants, timing their visits to the moment when garbage is being piled up at the end of the business day. Before dusk, they'll do one last pass of garbage bins in tourist areas before heading back to their nearby cliffs to sleep, bloated and spent. On more aggressive days, they'll raid schools, acting out their disdain for humans and their delight in dominating weaker animals.

Hamadryas are the only strictly patriarchal baboon species, and the silver-maned males that lead one-male units are ruthless in maintaining control through herding, biting, and visually intimidating the members of their female harems. But there are always challengers. Younger males regularly attempt full harem takeovers, which are tense and chaotic affairs that often involve a challenger kidnapping a rival's infant, which can end with the offspring injured or killed. After his installation, the new male's first diktat is then systematic infanticide, a cruel custom so embedded in hamadryas culture that females often spontaneously terminate pregnancies—an evolutionary adaptation, it seems, to avoid carrying to term an infant destined to be killed shortly after it is born. Meanwhile, the deposed rival, bereft of mates, will see its silver mane turn the same brown color as females.

This drama has been playing out across hamadryas societies for hundreds of thousands of years. In the Sarawat Mountains, though, the overwhelming supply of food available to synanthropic baboons has created a new social structure that has led to a reduction in aggressive competition. Driven by an overabundance of food, one-male units in and around the towns of al-Baha, Taif, and Mecca have ballooned to twenty or more baboons, far higher than the twelve or so typical of the species. That's left males here with a surfeit of mates, a situation that can, like a rapidly expanding midlife polycule, become too difficult to control. It's also introduced a modicum of female liberation into a society that's been rigidly patriarchal since its separation from other baboon species around 400,000 years ago. Hamadryas females within a troop of six hundred baboons were recently observed freely switching their allegiances across multiple units with no apparent reprisals and with no obvious fracturing of the larger social order. That's unlike hamadryas baboon societies in Africa, where the decision of a female to exit a unit portends its complete collapse and leads to the shunning of the previously dominant male, its weakness suddenly visible to all. Biologists have also observed a subclass of urban hamadryas females with no particular allegiance to any harem. Like orcas ramming boats and wearing salmon hats, this behavior is spreading rapidly across hamadryas society and becoming fixed in place—a true culture shift that might just portend a social revolution. These independent-minded females, dubbed "wandering females" by the Saudi government, have become powerful social drivers, deciding which one-male units to join and how much of their freedom of movement they will relinquish to the unit's male leader. That, in turn, has made females, rather than aggressive upstart males, the ultimate arbiters of the social order, and relegated violent takeovers of one-male units largely unnecessary, with a calming ripple effect across all of hamadryas society. Deep in this mountainous region, located in a kingdom notorious for its restrictions on human females, a female-led baboon culture is emerging with the power to reshape ancient customs.

Saudi Arabia is the designated protector of two of Islam's most holy sites, meaning human–baboon conflict isn't playing out solely as an urban drama but as a religious one as well. Mecca, around three hundred kilometers north of al-Baha, is the birthplace of the Prophet Muhammad and the site of the hajj, an annual pilgrimage that devout Muslims are expected to make at least once during their lifetime. Mecca is also in the home range of the kingdom's hamadryas baboons, and in recent years it's been the site of increasing incursions by the animals. Baboon troops numbering in the hundreds have descended on the highways that pilgrims use to enter the city, launching themselves into cars through open windows to scavenge for food and water. For a city so holy that only Muslims are allowed to enter, the presence of baboons is a black mark on Saudi Arabia's reputation as the keeper of the faith. As one foreign primate biologist who studied the kingdom's urban baboons put it to me, "It cannot be that in the holy city, baboons are shitting on the street." The baboons have become a fixture on Jabal al-Nour (Hill of Illumination), a small mountain on the outskirts of Mecca, where Muhammad received his first revelation of the Quran; the mountain attracts throngs of pilgrims, who must climb 1,750 steps to get to the top, and a growing number of baboons steal their water bottles and snacks along the way. Sometimes the insult goes too far. A video of a baboon stealing an *ihram* cloth, a ritual garment denoting purity, and dragging it across the rocky mountainside sent shock waves across the Muslim world.

It is this growing concern, I was told—that pilgrims to Mecca will be attacked and the city itself desecrated by animal feces—that prompted the Saudi government in February 2023 to launch a plan to "end the urban baboon problem" by 2026 after letting it fester for more than sixty years. The plan proposes a combination of population control through culling and sterilization, conflict de-escalation by minimizing open food sources, and habitat management by limiting urban sprawl into baboon home ranges. After letting the primates run roughshod over cities for so many generations, it's an uphill climb. Failure, though, would see the kingdom humiliated as millions of pilgrims fend off animals and sidestep feces on their way to communing with God.

■   ■   ■

It's 6:45 A.M., and I'm walking on a narrow al-Baha side road running alongside sheer cliffs, which reveal a view of the vast Sarawat Mountains as they tumble toward the Red Sea. On the way, I passed a couple sitting on a picnic blanket on a gravel shoulder, looking out at the view while a battered feral house cat, *Felis catus*, sat stone-still a foot away, patiently waiting for castoffs. The sudden drop gives me vertigo, making my head spin and my palms sweaty, but I can't look away. Ten yards down the cliff are a group of baboons tearing open plastic containers in the morning light. They frolic along the edge, then sneak up onto a cleared out vacant lot, where some climb upon a dilapidated steel chassis about the size of an SUV that looks like two crowd-control barriers stuck together. Beyond the chassis, another steel plank the size of a door rests at a low angle to the ground, propped up on some sort of debris. A female sits on top, staring passively beyond me. I count twelve in all: two males, hulking and silver, eight females, and two juveniles. A clan. One of the males sits in a heap, pulling apart a thin and transparent length of blue plastic. Another patrols the cliff edge on all fours, tail erect, its rump as red as pomegranate seeds.

By some unknown signal, the clan moves en masse, running farther along the cliff and up a short wall made of crude concrete bricks, poorly laid, with thin sticks of rebar sticking out on top, rusty red and dirty cream. Suddenly I'm reminded of the sunken detritus within which a giant Pacific octopus was living and dying in the polluted waters off Seattle, many worlds away. The baboons' movements are fluid, fast, and executed with a fixity of purpose that makes their status as predator undeniable. They walk along the wall for a few paces before disappearing through a narrow gap and out of sight. I watch them go, the last one a baby, screaming as he runs, his eyes wild, an infant cartoonishly aping his parents' intensity.

I approach the cliff edge to see them running at full speed down the near-vertical drop, a freedom of movement that I can't help but yearn for. About a hundred feet down they nestle together on a little outcrop, perfectly placed to stare out at the vast valley below. There, the clan merges with another, transforming into a band of about a hundred presided over by at least five or six large males, their spikey manes standing tall, like David Bowie's in *Labyrinth*. They keep their movements minimal, severe expressions on their dog-like faces, their long front arms ramrod straight as they patrol. The juveniles that surround them don't have that kind of control: They give in to their primal energy, hooting with their eyes bugged out, limbs slightly manic as they speed a few steps from the edge.

I walk back up to the road and under the shadow of a rock wall, following the concrete path up toward town. Suddenly I hear screaming—a short, weak call, almost like a sparrow's chirp—and scrambling across the wall is another hamadryas clan, dozens of baboons moving swiftly along the narrow escarpment toward some prize farther along the road. I stand there, dumbstruck, aware of the potential danger. They are a captivating bunch, with babies riding brown-tufted females, the clan led by a large male, who strides purposefully forward along the rock wall, another male patrolling from above it. They do not acknowledge me, or if they do, it isn't through a meta-communicative signal, friendly or hostile, that I can understand.

I turn away from the wall to watch them as they dip out of sight below the cliff's edge, leaving me with questions about what I am to them, if anything. Eventually, the juvenile's screaming subsides completely and I'm all alone, standing at the base of a small town in the middle of a mountain range teetering toward the sea at the edge of a massive desert. For the first time in a long time, I feel deep isolation, a primate without a clan, watching the dust from the footpaths of the monkeys settling onto the yellow ground.

I take a deep breath and turn around to continue up the road but immediately stop in my tracks. There, only a few feet from where I stand, three baboons are watching me from the high vantage point of the rock wall, having quietly taken up position while I was distracted by the flight of the disappearing clan. There are two females along with their male; he holds a dusty triangle of pita in his mouth as he stalks back and forth above me. Prolonged eye contact is a direct challenge among baboons so I do my best to look around instead of at them; I notice that they too are looking around me. We stay like this for a few seconds, the male patrolling, the pita dangling, the females looking alert and ready to take their cues. Is this what it feels like before an attack? And then, a sudden shift in energy and they all relax, scratch themselves, and collapse into a good impression of calm repose, their long tails dangling down the wall. I exhale, realizing only now how rigid my muscles are. The calm seems to have taken hold and I risk looking at them directly. For a second I catch the patriarch on the high wall looking me straight in the eye too. He's sitting cross-legged, scratching his leg, the half-eaten triangle of pita in his left hand, which he rests on his knee. Quickly, before I can register as a threat, I look away and turn to leave. None of the monkeys watch me go.

■ ■ ■

Synanthropes are defined by their relationship with human-modified environments more so than with humans themselves. That's an important distinction because, self-centered apes that we are, we like to think everything begins and ends with us. In cities, while we may be the most dominant species, we're certainly not the only one. We've created glorious, complicated, and messy ecosystems, set against the backdrop of healthy amounts of garbage, gum, and car exhaust. If you're the first mesopredator species to join humans in their habitats, as feral dogs are in this part of Saudi Arabia, there is only one rule: Find a way to live among humans. When you follow along in the footsteps of other species, the rules become more complicated. For hamadryas baboons, the task of colonizing urban centers like al-Baha has long been complicated by dogs, which have shielded humans here from the baboons' aggression. In the wild, baboons and canids frequently battle for territory and food sources. Packs of African wild dogs (*Lycaon pictus,* or "painted wolf"), a close relative of human-domesticated dogs (*Canis familiaris*), are in frequent conflict with resident baboon species. And while baboons are larger than African wild dogs and boast unmatched ferocity along with two-inch sharpened canines, packs of the canids actively prey upon their larger mesopredator rival, despite the risk of being killed or seriously injured. This dynamic has been transplanted into the artificial urban environment of al-Baha, with feral dogs retaining a hatred of baboons that runs so deeply in their biology that it will never be dislodged. Or at least that's what I believed until Ghanem al-Ghamdi tells me about the al-Baha dump.

One of the hundred or so baboon hot spots across al-Baha that his team identified, the waste management facility is in a class of its own. Where other sites harbor baboons in the hundreds, those that live amid the hills of refuse number, I am told, in the thousands. After a half-hour drive from the city center, we approach the edge of the dump, which is largely indistinguishable from the rest of the countryside. It's not a closed facility but a hilly area surrounded in parts by a low blue gate that stretches across the monotonous sepia-toned mountain landscape. From the road, there's little to see. But behind the barrier, nature's restless intelligence has fashioned cunning new variations of the universal order.

At a certain scale, all living animals cease to be themselves; they turn, instead, into a crowd. And, as Elias Canetti wrote in *Crowds and Power,* "As soon as a man has surrendered himself to the crowd, he ceases to fear its touch." Canetti was writing about how power is wielded to control and direct the masses, but the lessons he has for human organization apply just as well to baboons—though the entity that lives within the al-Baha dump is no mere crowd. Al-Ghamdi's estimates place them at a minimum of three thousand, though the true number could be twice as high. Left to their own devices, the hamadryas baboon masses that live here have collectively created a new social order that upends deeply rooted biological protocols.

Throughout the city of al-Baha proper, I saw baboons and dogs in a conflict that appeared as perpetual as nature itself. Both species are scavengers, are highly intelligent, move in groups, and rely on the same human food waste to sustain themselves. That makes interspecies violence a constant, with vicious daily bouts between dog packs and baboon clans to establish territorial dominance over the city's most prized locations. Even as the number of baboons has climbed, with bands of up to two hundred regularly descending on hot spots, small packs of dogs remain the first line of resistance against the city's primate incursion—a more reliable antagonist than humans, whose inconsistent efforts to remove baboons from their homes are regularly undermined by other people actively feeding them. "The male baboons," says al-Ghamdi, "they are tough, and sometimes they're aggressive with the dogs, but when a few of these bigger dogs get together, they can force the baboons to flee. Then they'll come back, especially the males, to defend their zone. This happens all over the city: It is an ongoing conflict."

But the dump has upended these rules. Just like in al-Baha proper and the plains of southern Africa, dogs share space with baboons here, but the rules governing behavior are dramatically different. Al-Ghamdi spent months observing the hundreds of dogs and thousands of baboons in this place. It was a natural experiment in the creation of a dystopian paradise: What happens to two ferocious scavengers, thrown into close proximity in a human-made space, when there are more than enough resources to go around? Not even thousands of baboons, it turns out, can consume the prodigious volume of food waste stacked in great plastic heaps amid these rugged hills. It is an endless supply, delivered daily, and its overabundance has all but zeroed out the need for violence between the species. But this dramatic reduction in inter-species aggression is just the beginning of the profound culture shifts occurring here. "It's an interesting thing. There are now …"—al-Ghamdi trails off, choosing his words carefully—"relations between baboons and dogs." I ask him to elaborate. "The dogs," he says, "have become submissive."

In the wide-open vista of al-Baha's waste management facility, at the end of the long chain of our elaborate global economy, hamadryas baboons have become the apex predator through sheer numbers, building a society in their image. These are typical primates, enmeshed in complex social structures; it is no wonder, then, that their instincts, when surrounded by other species, is to extend that social order out-ward, just like humans do. In al-Baha and at other dumps across the Sarawat Mountains, baboons have taken our lead, bringing *Canis familiaris* to heel.

The acculturation begins early, when the dogs are puppies and still reliant on their mothers for protection amid the chaos of the dump. At that point, an adult male baboon will crash into the family unit and kidnap one of its young, physically dominating the puppy by dragging it by its tail, sitting on it, and maintaining its physical separation from its mother, who, helpless in the face of the combined power of a one-male unit, is forced to watch the kidnapping, unable to intervene. The puppy mewls for help, terrified by the raw power and disinterested cruelty of the adult male baboon. There is only one lesson to be learned here: Survival requires total obedience. Any attempts at resis-tance will end with the death of the young. And so, as it has done for thousands of years with its other primate master, *Canis familiaris* learns to submit.

The kidnappings are the start of a process that ends with feral dogs becoming active members of baboon society. Once broken, the pup-pies will grow up amid a social order that submits to the silver-maned male baboons and their companions in the one-male unit. As they become physically larger, the dogs then play specialized roles as guards, keeping watch and playing with juvenile baboons, who will in return even groom the dogs as if they are a member of their own species. When there is an external threat to the baboon social order (which is most often other feral dogs), the dogs will serve their masters loyally and protect them at all costs.

I have seen few things more otherworldly than the image of a horde of baboons, thousands strong, moving in frantic unison over these desert hills. When it happens, triggered by onlookers, buzzing drones, or some other perceived threat, the dogs here move with the horde. Imagine thousands upon thousands of primates scampering up a steep hill, charging forward to flee perceived danger as if their lives depended upon it, the frenzied logic of the crowd having fully taken hold. In this melee, the dogs, dozens of them, instinctively space themselves out across the collective. Instead of running forward blindly, the canines maintain a modicum of calm, continually stopping and turning in their tracks to make sure that none of the monkeys is left behind, their herding instinct put to the service of baboon protection. If you squint, you can see individual baboons: large males dragging themselves forward on all fours, their stocky forms resembling silver balls rolling uphill; females, brown blurs, dashing at high speed, some with babies clinging to them; and juveniles racing to avoid the crush of their elders. At a distance, the fast-moving mob seems to move by a unified collective intelligence, enveloping an entire dusty beige hill in a matting of darker fur that then musters at its peak. Once its reached the summit, the collective disentangles itself just as quickly and scatters downhill at full speed toward its next point of safety, the dogs still herding, solicitous of their new pack, heads on a swivel, ears perked up.

Amid this commingling of monkey and dog, al-Ghamdi has also seen evidence of an even more bizarre phenomenon. He has studied this dump for years and has extensive drone footage demonstrating the many ways that feral dogs have become deeply integrated into hamadryas society. Recently, he has found evidence of a deeper connection evolving amid these endless trash heaps: Juvenile baboons have begun to ride the dogs. It's a disturbing and almost hallucinatory image out of context. For al-Ghamdi, it's evidence that the baboons are, as he puts it, "mastering the game." Some of the riding, he says, is nothing more than play, as juveniles are socialized to wrestle as a form of intimacy and relationship building. On top of that, an infant baboon quickly learns how to mount and dismount from their mother's back, an easily transferrable skill that would make riding a dog practically instinctive, especially given that the feral dogs of al-Baha are about the same size as an adult female hamadryas. However it has emerged, it is an undeniable sign that modern nature is alive in a landscape of refuse and decay.

Theories abound about what exactly it is we're seeing in the dump. Some observers believe that the kidnapping of puppies and their forceful domination by adult male baboons, coupled with the presence of adult dogs that not only guard troops but also submit to grooming, is evidence that baboons here are intentionally domesticating the resident canines. It could also be an inherent trait that they've extended to other species as a matter of instinct, as stealing infants is a deeply embedded behavior often used by males during one-male-unit takeovers. Hamadryas troops in Ethiopia have been observed stealing puppies and forcing the submission of feral dogs. Elsewhere in Africa, male baboons occasionally steal lion cubs and—rarely—even human babies. However, no one has yet tracked the entire life cycle of a stolen puppy in the al-Baha dump, so it's impossible to know whether these are brief muggings or sustained kidnappings that end with a kind of interspecies domestication. If it is the latter, this will represent the second time in the history of their species that *Canis familiaris* is being domesticated by a primate.

Regardless of how these behaviors emerged, they have become fixed in place by the unique and static conditions in the dump. Stasis is rare in natural ecosystems: rivers flow, trees grow and fall, storms stir up the land before leaving cumulonimbus clouds in their wake. Not so in the dump—or dumps, I should say, because al-Baha's waste management facility is not the only one across this region where feral dogs have become forcefully integrated into baboon society. The same culture shift has been observed in the Taif dump, about two hundred kilometers away—roughly twice the size of the maximum range of hamadryas baboons—making it unlikely that primates in one site learned the behaviors from the other, and more likely that both populations began adopting dogs independently. This suggests that a static urban landscape and endless food can give space and time for behaviors, like the occasional harassment of puppies by male baboons, to evolve into radical cultural movements that upend animal societies. It is not so far-fetched. After all, these same conditions—abundant resources and stable urban environments—are exactly what have allowed complex human culture to emerge and evolve at such high speed. And while baboons riding dogs might seem like a punch line, it may also mark a profound turning point in our understanding of the power of cities to reshape not only the human collective consciousness but also how other highly intelligent animals understand what it means to belong.

I can think of no better example of synanthropy in action: the feral descendants of domesticated creatures joining with urbanized primates to create a social order built on human trash.

Chapter 7

Deep Time

The fox could not build his own house, and so he
came to the house of his friend as a conqueror.

—Sumerian proverb, circa 2000 B.C.E.

Synanthropes Are Vectors. The animals that join us in cities are never alone. Within their bodies circulate pathogens that have found in their synanthropic hosts a direct line to human beings, and through us a pathway to replicate across the globe. The long history of our relationship with animals is inseparable from that of the viruses and the plagues that have sickened and killed us. At a time when the world is still grappling from the fallout of the last pandemic, the way we manage our relationships with synanthropes will decide how rapidly we will succumb to the next one.

■　■　■

We fear viruses, especially now that we know what they are capable of. It's only been a few years since the COVID-19 pandemic, when the best parts of human society—intimacy, connection, adventure, shared experiences—were abruptly taken away because a virus, against the odds, adapted itself to our bodies. Naturally, we think of viruses as blind automatons with only one function: attack. There's more to it than that. Viruses can make us sick and dead—short-term problems, from an evolutionary perspective—but they have also had more profound and lasting effects on how we have evolved. Overall, about 8 percent of the human genome is made up of the genetic remnants of previous viral infections, known as human endogenous retroviruses, or HERVs. These are like the photos of long-forgotten friends and neighbors that end up in the family photo album: They're not quite our biological family, but they've nevertheless become an inseparable part of our origin story.

The mechanism by which HERVs were integrated into our DNA is complex but boils down to the fact that viral genomes use many of the same basic components contained in our genetic code. Once these ancient viruses entered our cells, they interacted with our cellular DNA in ways that allowed them to become embedded there. Most HERVs do very little for us. Some, though, have become integral to our ability to survive and reproduce. A very special one, HERV-W, plays a critical role in maintaining human pregnancies by activating genes that allow for the efficient transportation of nutrients and oxygen from mother to fetus. Another, HERV-K, reduces the ability of HIV to replicate and infect human cells. Yet another, MER41, stimulates an immune response against flaviviruses like Zika and West Nile. Others do the same against viruses like measles, Ebola, avian flu, and coronaviruses. HERVs are ancient, the most recent one having been transposed into our genome 700,000 years ago, while others are the remnants of viral infections that happened as far back as 40 million years and became encoded in the genetic material of our long-extinct ancestors. Tens of millions of years before *Homo sapiens* separated itself from its most recent common ancestor (which was probably *Homo heidelbergensis*, an archaic human species that first appeared 800,000 years ago), viruses were already defining what it meant to be human. It wouldn't be the last time. As our species emerged out of prehistory and separated from the natural order, we once again became entangled with viruses, this time brought to us by a nascent class of wild animals—synanthropes—drawn to the ecological niches created by human ingenuity.

▪ ▪ ▪

The humans living in this place must have stared up at the openings in their roofs, which they used to pass in and out of their homes every day, and thought: *Will there ever be a way to use this door without inviting animals in?* The answer, of course, was no. And it's as true today as it was nine thousand years ago in Çatalhöyük, Anatolia, the world's first city. Built across the two shores of a riverbed, Çatalhöyük is a marvel of engineering. It's a honeycomb without roads or walkways, all of its hundreds of similarly sized homes sharing walls. To move across the city, people climbed ladders up through holes in their ceilings and strolled across the rooftops. It was also a hotbed of culture. There were polychrome wall paintings depicting geometric patterns and animals, beautifully constructed woven baskets, clay and plaster bas-relief, and dozens of sculptures and figurines, one of which depicts a Mother Goddess seated between two felines, which she appears to be stroking. Unlike contemporary human society, there was no nobility, no appreciable difference in the social status of the sexes, and no apparent political system other than collective decision-making. But just as now, there were synanthropes.

Çatalhöyük is one of the world's most important archaeological sites and has shed immense light on what society was like during the Neolithic Revolution, a period between 10,000 and 3,000 B.C.E. during which humans moved from small hunter-gatherer collectives into larger agricultural communities. And while there's much to learn from the site about emerging farming techniques, early goddess cults, and prehistoric family units, I can't help but be drawn to the fact that Çatalhöyük was, by all accounts, overrun by rodents—and not just any rodent but *Mus musculus,* the house mouse. It's the same mouse I hear every night before I go to sleep, scratching in the wall behind my bed as it runs its errands. Nine thousand years ago the inhabitants of Çatalhöyük were facing the same problems that I (and you, probably) contend with: animals in their kitchen. In one storage site in the ruins, bins of grains, seeds, almonds, peas, and wild mustard were surrounded by charred mouse pellets. It's evidence that, on the one hand, synanthropes have long found their way into our homes and, on the other, human progress, for all our self-congratulation, is still coming up short against the same problems that have bedeviled us for millennia. We also know, because of a curious practice whereby some citizens of Çatalhöyük were buried with animal scat placed on their torsos, that wild animals preyed on the house mouse and other small rodents in the urban center. It appears that residents would occasionally even invite these carnivores into their homes to rid them of their rodents. This was an early stage, scientists believe, in the self-domestication of felids that would later become our house cats, *Felis catus* (mine is called Beannie and she is seventeen years old). At Çatalhöyük, the African wild cat, *Felis silvestris lybica,* ancestor of all house cats, preyed on the small synanthropic rodents; eventually, it seems, the felids simply decided to live indoors, and humans abdicated to their demands. Remarkably, Çatalhöyük, despite being one of the first major, continuously inhabited, human urban centers, isn't the first to have a mouse problem. Far from it: *Mus musculus* became synanthropic more than twelve thousand years ago, at the bleeding edge of modern humanity's emergence. For the house mouse, Çatalhöyük was just the latest on a long, long list of welcoming human communities.

There were other residents too, far less benign than a mouse in your peas. DNA extracted from the bones of cattle at Çatalhöyük reveals the presence of tuberculosis, a bacterial disease, as early as 8500 B.C.E., which then spilled over into humans, confirmed via DNA extracted from the bones of a human infant. In 4500 B.C.E., there is evidence of another bacterial disease outbreak caused by salmonella, which is commonly transmitted by eating raw or undercooked poultry or eggs.

The people of the Battle Axe culture—a late Neolithic society dating from 2800 to 2300 B.C.E. that inhabited what is now southern Sweden and Finland, who revered their dead by burying them side by side with battle-axes close to their skulls—were also notable for being the first humans to die of *Yersinia pestis,* a bacterium carried by rodents. Four thousand years later, *Yersinia pestis* would cause the worst epidemic in human history: the Black Plague, which killed roughly half of Europe's population and as many as 200 million people worldwide, borne by fleas on *Rattus rattus,* the black rat, a synanthrope dating back to Neolithic India, whose spread across the world on the flanks of Mongol hordes changed the course of human history. And then there were the viruses.

The Neolithic Revolution was like a slingshot that suddenly launched our relationship with viruses into overdrive. Prior to the Neolithic, humans lived in small groups, usually of about twenty-five people, only rarely gathering in bands of about a hundred. That meant that if a virus managed to spill over into a human, there were too few transmission pathways for it to evolve enough to gain a foothold among our species and cause an epidemic. Because human groups were constantly on the move, synanthropes were also rare, so there was little sustained contact with wild animals, which is a prerequisite for viral spillover events. Once we stopped moving and built urban centers like Çatalhöyük, however, humanity found itself in the middle of an onslaught of pathogens. It was then that we first encountered Poxviridae, a deadly viral family that includes smallpox and mpox (the latter of which has its natural reservoir in rodents), hepatitis B virus (a close relative of which has its natural reservoir in bats), and measles (which diverged from a cattle-borne virus called rinderpest around 6000 B.C.E., at the height of the Neolithic), the most infectious virus our species has ever encountered. Permanent housing and agricultural surpluses allowed more of us to live together in larger and more complex social systems. That sedentary lifestyle and abundance of food attracted rodents and their predators into human population centers, which markedly increased the risk that spillover events would occur. For the first time since viruses emerged more than 4 billion years ago, the high density of people clustered in places like Çatalhöyük gave them enough of us to adapt to and kill. And finally, the domestication of livestock, including cattle, sheep, and goats, created multiple pathways for novel viruses to enter our bodies: They could be carried directly by synanthropes or be transmitted from them into livestock, which would then, through their saliva, breath, or blood, pass them on to us.

The expansion of human viruses during the Neolithic radically altered the trajectory of our species. But from our vantage point in the twenty-first century, we can see now that it was only the first of many acts. Medical anthropologists, looking back at our entwined history with diseases, have dubbed the Neolithic "the first epidemiologic transition." The second epidemiologic transition came during the first half of the twentieth century, when vaccination and antibiotics eradicated infectious diseases for much of the world's population (though, sadly, far from all). As we look ahead to a future dominated by artificial intelligence, scary weather, robot warfare, optional truth, dubious human longevity schemes, and the irrepressible energy of youth, we are now coming face-to-face with the third epidemiologic transition, when infectious diseases have returned with renewed intensity in the form of antibiotic-resistant pathogens and a wave of novel pathogenic viruses borne toward us with blinding efficiency by a growing army of synanthropes.

■ ■ ■

The Antarctic Peninsula unfurls itself toward the southwest edge of South America like a young fern, scattering a small archipelago of islands across an ice sheet thousands of kilometers long. Along this coiled and narrow stretch of land, the high peaks of the Antarctandes mountain range impose themselves upon the sky, covered by a rough latticework of white snow against black rock face that resembles break lines creeping across a smashed windshield of tinted glass. It's cold, dry, and windy, a landscape unfit for human habitation. But it's a perfect place for penguins.

"The penguins," says Victor Neira, "don't see us as predators. They approach you with curiosity and don't run away. So we capture them quite easily with a fishing net." Neira is a veterinary scientist who travels regularly to the Antarctic Peninsula to study many of the continent's local animal populations. It's evident that penguins are his favorite. "You just put the net over them," he continues, "and pick up the penguin like it's an American football, tucked below your arm, with its head behind you." Neira's words are punctuated with violent coughs, for which he apologizes profusely; he is sick with COVID-19, he tells me, his voice tinny over the phone line. As an expert in viral transmission across animal populations, it's fitting that his enthusiasm never wavers despite the virus coursing through his system. After nabbing a penguin, Neira says, the next step is to extract a sample, which is easy enough if you know what you're doing. You quickly swab the penguin's cloaca (a single opening for everything inside the body that must come out, including eggs) or take blood from the veins on its rubbery feet. To make sure the bird stays calm, Neira explains, "you usually put something like a sock on its head to stop it from seeing," which keeps it immobile. "But also the most important things are the wings," he adds, "because if you let go of the wings, they're going to hit you very, very, very hard. Oh, and also it will peck at you with the beak." The pecks, flaps, and squawks are all worth it for the samples, which Neira uses to track the prevalence and trajectory of viruses across this remote region. Antarctica's distance, uncompromising climate, and breeding ground for migratory animals like penguins and residents like seals, make it an ideal sentinel site for the rest of the world. If a virus is circulating here, the thinking goes, it's only a matter of time before it will make its way everywhere else.

Neira projects an easy calm. Perhaps it's the COVID exhaustion, though it might be because he's talking to me over a cell phone from his office at the University of Chile in Santiago instead of via satphone from the icy wastelands of the Antarctic Peninsula, a place that requires constant fortitude. To reach the penguin colonies that inhabit the edges of the ice sheets, Neira and a team of three other scientists and graduate students take a fifteen-hundred-kilometer journey by boat (two to three days through rough seas) or military aircraft (around five hours) from Chile to Base Presidente Eduardo Frei Montalva, a research station on King George Island, a tiny mound of rock just north of the Antarctic mainland. From there, Neira and his team are driven by snowmobile into an even more remote area on the edge of the Antarctic ice sheet and left to do the work. "They leave you in a cabin with some canned food, and you have to stay there for two weeks or a month, just three or four people." It's an unforgiving landscape, prone to windstorms, fierce waves, and sudden drops in temperature to as low as −4 degrees Fahrenheit. Given that scientists aren't known for their survival instincts, the group, Neira says, always includes one survival expert.

Neira only began his work of sampling penguins in Antarctica ten years ago, but it was a completely different era. This was 2015, four years before COVID-19, when few of us could fathom that a global pandemic was possible in our lifetime. It was also a period when the viruses circulating among penguins were generally benign, the most serious of which was a well-documented influenza strain endemic among the birds. "We were studying the regular influenza for years," Neira says. "That virus doesn't kill the animals. So we would go from the research station by helicopter or Zodiac boat, and work for four to six hours before heading back to the base." Despite the extreme environment, the work itself was straightforward. "You could collect samples from the animals if they looked healthy without major protection," he says, referring to biohazard containment protocols. "Even when the animals were dead, we collected the samples but just had to be a bit more careful." That meant using gloves when handling carcasses but little else in the way of precautions.

All that changed in 2023. "The highly path is different . . . the highly path kills the animals." Neira is referring to highly pathogenic avian influenza, also known as H5N1, a virulent flu strain first identified in Hong Kong in 1996 that has since been spreading and evolving within animal populations the world over. During the past thirty years, while tens of millions of animals have died, less than a thousand human cases have been recorded. Troublingly, though, the human mortality rate is a staggering 50 percent.

When Neira detected the virus's arrival in Antarctica, it was an ominous sign. In his many years of research missions, it was rare to see dead animals other than the very young or very old, most of which presented as single corpses left alone on the ice. But suddenly the dead were congregating. On his way from a remote cabin to a penguin colony, Neira and his team came across the corpses of six skuas, predatory birds resembling darkened seagull with sharpened beaks. Known to harass and eat penguin chicks, as well as scavenge off the remains of larger animals like seals and sea lions, skuas feed opportunistically on whatever animal flesh they can find—making them a perfect conduit for a highly transmissible virus like H5N1, which efficiently spills over across species. The birds were the most plentiful victims of the virus, but they weren't the only ones. Soon, Neira came across the desiccated bodies of Antarctic fur seals, their flippers flattened against the ground, some lying twisted in a half-moon shape a few steps from the ocean's edge.

The arrival of H5N1 signaled that the Antarctic world had changed, and so too would Neira's. Gone were the unprotected forays into penguin colonies. Before his research missions, against the solemn backdrop of the cloudless Antarctic sky, Neira and his team now readied the ritual weaponry they would need to protect themselves from the scourge: white hooded polypropylene coveralls, KN95 masks, blue nitrile gloves, boot covers, and black goggles—the full complement of PPE, stark and antiseptic against the polar landscape, making them look like cosmonauts navigating an alien world, armed with syringes as they advanced upon the dead.

For Neira, none of this was new. That's why it was so terrifying. When he first detected H5N1 among Arctic seals in 2023, it was a few short months after a viral epidemic had laid waste to Chile, his home country, leaving tens of thousands dead. As he braved the ice fields, images of that apocalypse—the bodies, the mass graves, the stench—passed through his mind. This time, he hoped, he wouldn't be too late.

■ ■ ■

There have been more epidemics caused by novel viruses in the first two decades of the twenty-first century than in the entirety of the twentieth century. Something has shifted that has made us more susceptible to infection than, perhaps, ever before. This is the third epidemiologic transition, and it means that we are facing two hard truths. First, the COVID-19 pandemic wasn't a one-off event. And second, the conditions that led to it are intensifying, even as we continue to largely ignore this very hard problem. Yet instead of holding both truths, we've become fixated on a seemingly binary choice: believing that SARS-CoV-2, the virus responsible for the COVID-19 pandemic, was transmitted from an animal host or that it was the product of a lab leak from the Wuhan Institute of Virology. It's an important question, arising from the understandable anger about our inability to conclusively track the origins of the pandemic. Still, it misses the larger point. SARS-CoV-2 (2 percent mortality rate) wasn't the first or even the second coronavirus to emerge and sicken us in the twenty-first century. It was the third, following quickly after SARS (severe acute respiratory syndrome, 10 percent mortality rate) in 2002 and MERS (Middle East respiratory syndrome, 35 percent mortality rate) in 2012.

Epidemiologists like me are open to using cross-sectional data to help explain how epidemics arise. But there's nothing like a pattern to clarify where the risk lies—and where things are headed. The emergence of just one deadly coronavirus, SARS, after the viral family was written off as largely benign, is a curious anomaly. Three coronaviruses in two decades—a period that is, on the scale of viral evolution, just a blink of an eye—is a series of blaring alarms warning us that smoke isn't just rising in the distance; the building we are living in is engulfed in flames.

SARS-CoV-2 isn't even the end of it. Two more never-before-seen coronaviruses were detected at the height of the COVID-19 pandemic: a porcine coronavirus that sickened schoolchildren in Haiti and a canine coronavirus that caused a cluster of pneumonias among rural children exposed to wildlife in Malaysia. Add to that the sudden emergence of a Zika epidemic in 2015, along with novel strains of dengue, yellow fever, norovirus, and H5N1, and it's clear that our relationship with viruses is being rewritten at hyper-speed. Simply put, the risk of a spillover has been so ratcheted up that many virologists now believe we have entered a new pandemic age, when these massive global calamities will inevitably arise with ever greater frequency. And synanthropes are playing a leading role in this grand drama.

Chiropterans have the dual distinction of being the world's most successful order of vertebrate synanthropes, while also being among the most well adapted to harboring a grand virome of endlessly mutating viral strains capable of infecting humans. These include rabies (close to 100 percent mortality rate), Marburg (80 percent mortality rate), Nipah (70 percent mortality rate), and Hendra (60 percent mortality rate), along with a seemingly inexhaustible supply of coronaviruses. Because they are so lethal, it is highly unlikely that these viruses could ever evolve within human hosts to the point that they become benign; they simply kill us too fast. But in bats they are the product of continuous transmission chains over virological deep time, allowing them to proliferate endlessly within colonies around the world, priming them to spill over when the moment allows.

Synanthropic bats have long populated human settlements, and the direct part they play in viral spillover events is well-known. In Bangladesh, for example, raw date palm sap is an ancient staple, but the viscous gray-yellow liquid is also a favorite of synanthropic bat species.

In villages where the palms grow, collectors shave the tree trunks to expose the sap-wet wood inside, then collect it as it drips down. Indian flying foxes—a species of megabats with faces that resemble small dogs, and which are a natural reservoir of Nipah virus—are drawn to the sugary sweetness, alighting nimbly on the palm trees and licking the sap as it runs down their mottled brown trunks. They also piss and shit while gorging themselves, their excrement intermingling with the sap as it runs down the taps shunted into the wood, which ferry the liquid into clay pots suspended below. This transforms the raw date palm sap, a delicacy in this region, into a massive inoculum dose of Nipah virus, for which no vaccine exists.

The transmission dynamics of Nipah are an easy problem for epidemiologists to untangle because there is no intermediate animal host. Infection happens in a straight line from one mammal to the next, so the only mystery is how the virus gets from the body of the bat into ours. In human-modified habitats replete with many different synanthropes, transmission dynamics splinter because it's impossible to predict how a multitude of species adapted to life with humans—bats and raccoons, civets and pangolins, wild boar and white-tailed deer, feral dogs and hamadryas baboons—will adapt to one another. You may think that urban coyotes would feast on the badgers they find in parks and gardens and catch rabies as a result. Instead, the two species just as often end up hunting together, the badger shaking underground prey loose, the coyote chasing it down when it tries to flee, both ending the day well-fed. Until I went to al-Baha, I would have never thought that baboons and feral dogs could share in a social system that ends with the unthinkable spectacle of the dogs submitting themselves to primate riders, a situation that could turn the canids into vectors of rabies, tuberculosis, and streptococcus. Red-tailed hawks might reduce feral pigeon populations by hunting on the wing, or they might cause them to explode by instead culling rodents that compete with the winged scavengers for scraps of food; the result could be a greater risk of either salmonella spread from rats or avian flu from pigeons. The intersections between synanthropic species, pathogens, urban spaces, and our own fragile immunity are as complex as an octopus's skin. The transmission pathways are practically infinite, making the challenges of predicting them exponentially harder. Add to that viruses that have honed their capacity to transmit across species and kill with precision, and the solidity of the world melts away. In its place? The new pandemic age.

▪ ▪ ▪

Chile is one of the world's longest countries, with a coastline over six thousand kilometers long and an average width less than two hundred kilometers, running along the southwest edge of South America. Naturally, Neira, like three-quarters of Chileans, lives near the coast, making him one of 15 million humans who regularly interact with South America's largest population of marine mammals. The South American sea lion, *Otaria flavescens* (yellow-eared seal), a pinniped that can weigh close to eight hundred pounds and has the look of a straw-colored Bernese mountain dog, is a fixture of life in Chile, where its population reaches into the hundreds of thousands. They're an impossible-to-miss synanthrope in the country, where pretty much everyone's had some kind of run-in with the barking beasts. "As a Chilean, you can easily see sea lions," says Neira. "You can buy fish heads and feed them like that." Aside from daily handouts, sea lions near the capital city of Santiago have also set up their colonies near local fish processing plants, where they gorge on fish guts, bycatch, and other detritus dumped back into the ocean by commercial fisheries. As a veterinary scientist and expert in virology, Neira has had more than his share of interactions with the pinnipeds. "They're not aggressive by nature," he explains. "They see people all the time, so they generally just move out of the way for you." Neira thinks for a second and seems to change his mind. "That being said, if you mingle with them, they will try to bite you. So usually you have to kind of run or you have to scream."

"Scream?" I ask.

Neira laughs. "Yeah," he says, before letting out a bloodcurdling cry. I ask if he's ever been bitten. "No, no, no, but I've been very close. If you run . . . Well, the sea lions are very fast." He doesn't finish the thought, but it seems Neira is an even faster runner. "One sea lion tried to bite my coworker in the feet," he adds, "but he had good boots, so nothing happened."

Neira's is multifaceted work, equal parts field research and laboratory analysis, with much of his time in the lab spent sequencing viral samples he's collected from seabirds and marine mammals. His work is nearly constant because the more samples he and his team collect, the more likely they are to detect novel viruses or variants when they first arrive in the region. This is the epidemiologic research, undertaken in countries across the world, that makes up a network of scientific early-warning systems designed to give humanity a head start in identifying novel virus outbreaks and tamping them out before they explode into the next pandemic. Sometimes, no matter what they do, there is just no way to stop the inevitable. In December 2022, after spending years watching warily as H5N1 tore a destructive path through animal populations the world over, Neira watched helplessly as the virus finally arrived in full force in his home country.

The first cluster of highly pathogenic avian flu cases Neira discovered in Chile was among pelicans, cormorants, and terns, three bird species that regularly cross paths with other avians, sea creatures, and the viruses they harbor along their vast migration routes. (In the case of the Arctic tern, "vast" is no hyperbole: It has the longest migration route of any animal, from the Arctic Circle to Antarctica every year, a distance of 25,000 miles.) "When the virus first came to Chile, the viral genomes we sequenced in pelicans and other seabirds were practically identical to samples identified in Peru, and those showed that the virus had originally come from the United States." Having seen the damage that H5N1 had already wrought, with tens of millions of wild birds killed, and hundreds of millions of poultry culled to contain its spread, Neira fully expected to see an outbreak among the country's various bird populations, but what happened next shocked him.

Within a month, after clusters of dead seabirds began washing up on shore, Neira's team began observing cases among another resident species: sea lions. "We just found the dead animals in the sea and on the shore," he says, "with some showing signs of a respiratory virus." Within days, what started as a cluster of cases ballooned into something else. After years during which the monthly number of sea lions stranded on land never surpassed 100, January 2023 saw a sudden spike to about 250. In February, that number doubled to 500, and the numbers continued to rise precipitously, with more than 4,500 sea lions stranded in June 2023 out of a total population of 130,000. Within six months of the first H5N1 case among Chilean seabirds, more than 30,000 sea lions had died, with the total number across all animals—including birds, seals, dolphins, and even humpback and fin whales—reaching an astounding 120,000 deaths, the interspecies epidemic stretching along all six thousand kilometers of Chile's coastline. Overall, Neira estimates that at least 10 percent of the sea lion population in the country died in that six-month span. It was a sudden and deeply terrifying mass mortality event caused by a virus that had suddenly gained the ability to transmit and kill its mammalian hosts.

When tens of thousands of animals weighing half a ton or more are killed within a few short months, the challenge quickly moves from the epidemiologic to the practical. "It was very problematic for the affected communities," Neira says, "because the sea lions are very heavy, so the dead animals on the shore became a big problem. They had to bury them quickly to try to avoid infection, flies, and other scavengers finding the bodies. And all the animals were basically dying at the same time." At first dead animals were being placed in large freezers, but that became unworkable as the number of the dead rose precipitously. Soon, the government was putting up makeshift buildings to store the animal remains. After a few months, those buildings were replaced by large, more permanent halls. Because of their massive girth, the sea lion corpses couldn't be hauled by hand, forcing the Chilean government to raise a fleet of Caterpillar tractors, which they used to dig trenches, lift the dead animals, and deposit them into mass graves. Some of the gravesites were hastily dug into beaches, steps from the shoreline, an efficient yet unceremonious reminder of how cheap life can become when everything around you is dying.

No country escaped COVID-19, but most of us were spared its true horrors. Forced indoors by shelter-in-place restrictions, the scale of the tragedy still remains largely abstract. Close to a million people in North America and almost 10 million died worldwide, but the visceral horror of seeing and touching the bodies of our loved ones was replaced by a set of graphs charting the rising numbers of the dead. Many of us simply never saw the pandemic's victims, leaving us unable to truly reckon with its legacy. In Chile, though, there was no avoiding the outsized toll of H5N1. Mass death was visible everywhere, in the carcasses of dolphins and fin whales, sea lions and seagulls, and penguins and pelicans, strewn haphazardly at the water's edge. Along thousands of kilometers of coastline, Chileans witnessed the slow corrosion of animal flesh by sand and saline and its rapid ingestion by mesopredator scavengers and insect opportunists. The entire environment, as far as the eye can see and farther still, transformed into a memento mori, the only movement left being the endless empty tide.

■　■　■

Viral time is measured by molecular clock analysis, a way to estimate how long it's been since two closely related viruses branched apart. It's done by tracking the endless tick-tocking of nucleotide bases in viral RNA or DNA as they randomly mutate during replication, thereby reducing or increasing a virus's fitness to infect, replicate, and transmit itself across host populations. This technique has been used to estimate when the lineage leading to SARS-CoV-2 first branched off from the group of SARS-like bat coronaviruses that spawned it (in 1982), and to measure when HIV might have first emerged (between 1881 and 1918, with the first spillover event occurring in rural southeast Cameroon sometime around 1920). It's shocking when millions of years of viral evolution suddenly intersect with the few paltry years of our own existence. How could a barely living pathogen find its way into us at the exact moment when it has randomly evolved to take advantage of our biological vulnerabilities? The chances are infinitesimally small; and yet, Earth's virome—the totality of all virus particles—is astronomically large: There are 10^{31}, or 10 nonillion, viruses on our planet, outnumbering the stars that light the universe. Still, how can it happen?

This is the story of SARS-CoV-2 and the pandemic that followed, with its sickness and its death, along with the collective courage and the temporary insanity it brought to our species. Though the exact path the virus took to spill over into humans remains unknown, the virus's genome—all of its roughly 30,000 nucleotide bases—has been mapped in exquisite detail. And by using molecular clock analysis, like following a string back in time through a virus's labyrinthine mutations, we know that the closest ancestors to SARS-CoV-2 were viruses carried by the Rhinolophidae family, commonly known as horseshoe bats, a long-standing and successful synanthrope with niches the world over. But that's not the whole story. As scientists have sought to reconstruct the exact viral strain and date that started the COVID-19 pandemic, the numbers have failed to add up. The problem is that almost immediately after the first case was detected in November 2019, multiple variants of the virus were already circulating, which should have been impossible based on the date of the first detected strain. As they dug further, the answer soon came into focus: SARS-CoV-2 could not have been caused by a single transmission event. Instead, the virus gained a foothold in our species through multiple zoonotic spillovers, likely in the Huanan market in Wuhan, during November and December 2019. These repeated spillover events gave the virus multiple attempts to spread among humans, meaning that even if one of the variants failed to take hold, others were ready to test their fitness in the human respiratory system. The proof of this grand strategy, operating at the ruthless and unconscious level of natural selection, was that three distinct viral lineages of SARS-CoV-2 crossed over into humans in rapid succession. This speaks to the variegated paths that each took to find its way from bats to humans, which almost certainly involved passing through other synanthropic mammals that regularly come into contact with both bats and humans. These detours forced the viruses that would become SARS-CoV-2 to mutate or perish in the strange new respiratory systems they found themselves in—a final push that transformed them into efficient and deadly transmitters.

■　■　■

Buried within the synanthrope-driven tragedy of the pandemic, a curious counternarrative played itself out. It took billions of years for our species to intersect with the virus, but it took only a few short months for us to retreat from the world and let animals claim for themselves the bewildering landscapes we created. Within five days of a lockdown in Wales, a herd of Kashmiri mountain goats, glossy white wool hanging loosely down like vines, their heads adorned with dark curved horns, moved into the seaside resort town of Llandudno, where they clambered up short retaining walls, chewed on hedges, and walked unhurriedly through empty streets. In Mumbai, India, a megalopolis of 15 million people, frilled indigo-colored peacocks wandered out of a nearby forest one week after a countrywide lockdown, spreading their plumage in a display that took up multiple lanes on an emptied road, their feathers shimmering like dozens of eyes surveying the scene, now absent of humans. In Los Angeles, P-22 became an international celebrity as the mountain lion wandered out of Griffith Park and into the empty neighborhoods that surrounded it, a stealthy and beautiful vision of freedom. These wild animals parading through the hearts of cities were cast as miracles, but they were really just synanthropy in action. P-22 had relied on big game like mule deer to survive, but few were available to him in Griffith Park, forcing him to make incursions farther afield in pursuit of small rodents. In Mumbai, the peacocks had been sequestered within a nature reserve, enclosed for their own protection as the city swelled on all sides, effectively turning them into unwitting captives, locked in their natural habitat with the city walls closing in. And the Kashmiri goats of Llandudno, a herd descended from a pair given to Queen Elizabeth by the Shah of Persia in 1837, were normally residents of the nearby Great Orme, a large outcrop of rock and short grass. In the years preceding the lockdowns, they had been venturing steadily farther into town to flee increasingly violent storms brought about by climate change, preferring the protection afforded by city gardens and the retaining walls that surround them to the large and open steppe-like crag. It wasn't the pandemic, then, that brought these animals out of the shadows. COVID-19 simply gave a last push to desperate species seeking refuge from the violent perturbations of the Anthropocene.

▪ ▪ ▪

The world is set up so that everything is at hand. It's a seductive feeling. No matter where you find yourself, there is a fulfillment center nearby, a gig worker on an electric motorcycle—hell, even a drone that will deliver to you the things you desire so that you don't have to look a stranger in the eye. We get to star in our own story, the world organized around our needs, goods and services orbiting our bodies like planets around a gaseous star.

It's not just in your head. Global markets have led to supply chains that connect rare earth minerals excavated from a mine in Longnan, China, to Foxconn maquiladoras sitting on the high desert mesas of

Tijuana, Mexico, to an Apple Store in Grand Rapids, Michigan. Aesop Rock put it best: "You can buy a Jet Ski from a cell phone on a jumbo jet; T-E-C-H-N-O-L-O-G-Y, it's the ultimate." No sector is immune, the magnetism of complex economies too powerful. Not even animals can escape its draw.

The global meat industry is worth $1.8 trillion annually, a number that's expected to rise by 50 percent within the next decade. Like other sectors, it's all about efficiency. In the case of animals, that means bundling them together as closely as possible to maximize the amount of meat that can be produced by square footage of grazing land—or, at its extremes, by cubic footage of factory farms. Domesticated animals are not synanthropes. But we feed on them and their milk and their eggs, and this deep intimacy makes them efficient in another way: as viral vectors within which new pathogens can evolve and find easy access to our bodies. While domesticated animals, so well contained, are unlikely to be the source of novel viruses, if they are infected with one, they amplify the risk of it finding its way to us. Combine tightly packed animal herds with the crush of synanthropes bringing viruses from wild places into cities, and it's inevitable that outbreaks, epidemics, and even pandemics will continue to threaten us.

Scientists now recognize that among the greatest sources of novel pathogens are casual synanthropes: animals that spend part of their time in human-modified environments and the rest in the wild. The H5N1-fueled mass mortality event among Chilean sea lions is a perfect example of how casual synanthropes can ratchet up pandemic risk. I've argued in this book that cities are islands, in more ways than one. They're urban heat islands, with microclimates that behave differently from the areas that surround them, islands of safety and nutritional wealth that force animals to stay in place, and genetic islands of dwindling alleles that might end up leading to the extinction of urban species or the birth of new ones. But from a virus's perspective, cities are nodes within a network, replete with massive populations vulnerable to infection. And for that reason, they're where outbreaks are most likely to transform into epidemics, or worse. And just as the threads of our economic interconnectivity have never been more complex, viral pathogens can now be efficiently borne from the most remote places on Earth into the heart of human metropolises in no time at all. It just takes an animal to bring it from there to here.

H5N1 is often dubbed avian flu, but the virus most likely to spark the next pandemic is in fact circulating in all kinds of animals, including dairy cows across the United States. As of June 2025, more than one thousand dairy herds (4 percent of all U.S. herds) were infected across seventeen states. It's understood that the virus was passed from migratory birds, which entered dairy farms, ate feed, and drank water, exposing cows in the process. But from there, it gets murky. Dead pigeons, blackbirds, and grackles were found on the index farm where the virus was first detected, implying that these avian synanthropes could have acted as intermediate hosts, ferrying the virus from their wild cousins to cows. In one Texas dairy farm where a human being became infected, barn cats tested positive for the virus, revealing a whole new synanthropic pathway the virus exploited.

Advances in genomic sequencing have allowed scientists to peer into the evolution of the avian flu virus in real time to understand how long it might take for it to mutate into a strain capable of rapidly spreading among humans. The news is troubling. How close are we? In December 2024, researchers at Scripps Research Institute at the University of California, San Diego, discovered that a single mutation in the most prevalent H5N1 strain, known as 2.3.4.4b, would shift it from attacking receptors in dairy cows to those found in humans. Worse, as an RNA virus, H5N1 is prone to many mutations every time it replicates. It's a change that could occur at any moment, among any single virion, anywhere in the world. After that, the researchers say, H5N1 would be transformed into a pathogen capable of starting the next pandemic.

▪ ▪ ▪

The mass mortality event in Chile is a startling reminder of the power of viruses to bring about apocalypses. But even though tens of thousands of sea lions infected with a mammal-adapted virus died within a few kilometers of 15 million human inhabitants, the number of human cases during the six-month epidemic was remarkably low. In fact, Neira says, only one person became infected in the entire country. Given how close H5N1 is to evolving for human transmission, as well as the number of opportunities for spillover the virus was afforded on account of its proximity of Chile's sea lion and human populations, it feels like a lucky break. But it's not the epidemic's only mystery.

"The mass mortality events went up from January until June 2023," Neira says, "but after that, no more cases were observed among sea lions." Since August 2023, despite regularly sampling the pinnipeds, Neira's team hasn't found any traces of the deadly virus among sea lions on the Chilean coast. H5N1, Neira explains, was simply too efficient. "It killed off all of the susceptible animals," he says, "and the ones that survived are either resistant to the virus or they weren't exposed."

I ask Neira about his personal experience witnessing the episode of mass death.

"I didn't actually observe any myself," he says with a laugh, "because the population of sea lions where I live, in the capital, was completely unaffected by the virus."

"Oh!" I exclaim. "That's lucky."

"Very lucky," Neira agrees. "But it's more than that. The sea lions in the city are very, very focused on the leftovers from the commercial fishermen, so they never leave the area to try to capture fish. Because they have the food here already, they didn't interact with migratory seabirds out in the ocean, so they were never exposed to the virus." It's an unexpected twist. While casual synanthropes are at the bleeding edge of viral outbreaks, it appears that, in Chile at least, the more fully synanthropic a species is, the more protected it might be from novel epidemics. That would make cities something other than transmission nodes for viruses as they crisscross the globe. Sometimes, it seems, an animal's fierce loyalty to living in an urban island might just be what saves it from annihilation.

▪ ▪ ▪

The emergence of avian flu is a textbook case in the interconnectedness of the planet, and a cautionary tale (hopefully no more than that) in the need to recognize the signals emanating from the natural world when the risk of a pandemic accelerates. While those signals have been known for quite some time, it was only in 2004, two years after the first SARS epidemic, that a group of concerned epidemiologists, virologists, veterinarians, and ecologists codified a strategy for pandemic prevention that placed humans, animals, and the environment within a single continuum. They called it "One World, One Health," and since then, One Health has become the consensus approach in how to understand and prevent pathogenic threats. It's also intuitive: If we are to survive, according to One Health, we cannot separate environmental health, animal health, and human health. One of the lessons here is that urban sprawl can cause wild animals to become synanthropic (think about the Lincoln coyote, trapped within the expanding urban matrix and forced into contact with humans and their pets), which can ratchet up our risk of infection from the novel viruses they harbor. Another lesson is that climate change won't just intensify cataclysmic weather events but will irrevocably alter the global movements of animals and their pathogens, which could bring them both into closer contact with our species.

One Health allows us to link the years-long epidemic of H5N1 in North American cattle herds directly back to habitat destruction, which has forced infected wild birds into contact with domesticated animals. This same strain, Victor Neira and his team discovered, borne by migratory seabirds, subsequently ended up killing tens of thousands of sea lions in Chile, half a world away. These are terrifying reminders of our continuity with the world at large and illustrative of why One Health is such a powerful approach to preventing pandemics before they occur.

Still, there are reasons for hope. In the wake of COVID-19, humanity is far better prepared to weather and even avert the next pandemic. The global network of scientists like Victor Neira is able to rapidly send up flares once novel viruses are detected, providing us with a chance to get cures—like antiviral medications and vaccines—to places that urgently need them. That assumes, of course, humanity is able to create vaccines for novel viruses fast enough to deploy them before outbreaks metastasize into pandemics. Up until now, though, vaccines have been created only after a pathogen emerged that did us harm. But with the third epidemiologic transition in full swing, that backward-facing strategy just won't cut it anymore.

The new challenge facing us is so hard that it sounds like a paradox: How do we create a vaccine for a virus that doesn't yet exist? The answer is to shift our focus away from specific viruses and toward the broader planetary virome. Of the 368 viral families known to exist, only 26 have produced strains that can infect humans. While there's no way to predict what the next pandemic-causing virus will be, it's a pretty sure bet that it will come from one of these viral families. What's more, the new strain will almost certainly resemble existing strains that have infected us before, because new viruses must necessarily evolve out of existing ones. Rather than trying to develop strain-specific vaccines, then, virologists have shifted to assembling vaccines that simultaneously target entire families with a track record of infecting our species. The thinking is that if you build a vaccine out of components from a family's existing strains, you'll have a good shot at eliciting partial immunity to whatever new deadly virus emerges next because it will likely include components from a strain that already exists. The effort has already begun with coronaviruses: Bart Haynes and Ralph Baric, two leading coronavirologists, are developing vaccine candidates that mix and match components from twelve different group 2 coronaviruses, including SARS, SARS-CoV-2, and their closest relatives, along with MERS. In animal studies, they've found that a single multi-target vaccine candidate can protect mice from all of the group 2 coronaviruses known to exist, and for that reason will almost certainly provide some protection against future coronavirus pathogens too. Other teams have done the same for seasonal flu, creating a blended vaccine that is effective in staving off infection from the unpredictable grab bag of seasonal strains that typically make us sick.

These are cutting-edge technologies that could be used to contain outbreaks through a strategy called ring vaccination, in which frontline health workers and the close contacts of people infected with a novel virus are rapidly vaccinated, which would shut down transmission pathways the pathogen needs to create a human reservoir. Ring vaccination has proven effective at ending Ebola and smallpox outbreaks, but the key is speed—that's why we need broad-based vaccines for every viral family that can infect humans if we're to seriously plan on avoiding the next pandemic. The scientific challenge, it seems, is only one hurdle. For the past fifty years at least, scientific innovation has overwhelmingly focused on incremental improvements to existing cures rather than on true leaps forward, a product of market forces that prioritize de-risking future profit at the cost of meaningful technological advances. That regularly nets us more potent Tylenol formulations or slightly less nauseating chemotherapies but gets us no closer to creating vaccines for potent viral enemies capable of doing us harm.

The most recent pandemic is still with us, despite the fact that everyone wants to forget it. We still feel its malaise and the fractures it has riven across social groups and societies. Still, we are far better prepared now for the third epidemiologic transition than before the emergence of a cluster of pneumonia cases in Wuhan in November 2019. And it's not just the vaccines. In the wake of COVID-19, One Health has become the consensus approach to understanding how and where to look to prevent pandemics. In March 2022, four United Nations agencies, including the World Health Organization, the Food and Agriculture Organization, the World Organisation for Animal Health, and the UN Environment Programme formalized a One Health Action Plan to eliminate pandemic threats. It starts with a reckoning: We cannot protect humanity if we don't recognize that the health of animals and the way we direct their movements through our communities will decide our collective fate more than any other action. And it sounds like scientists, at least, are getting the message. Before the pandemic, there were barely seven hundred peer-reviewed studies published per year about One Health. Since then, the numbers continue to rise, with more than five thousand studies published last year.

■ ■ ■

Despite the sudden and near-miraculous disappearance of H5N1 from Chile's wild animal populations, Victor Neira remains resolute. "It was strange to see dead seals in the Antarctic," he says, "and before 2022, we didn't observe a single case of the highly path among them." But recently, one of his students, dressed in the now-standard full-body PPE, sawed open the skull of one old and desiccated seal corpse found at the edge of the Southern Ocean. When Neira analyzed the brain tissue, it was positive for H5N1, and it wasn't the only one: Neira's team is now regularly finding the virus in Antarctic animals. "To follow the virus—this highly path—this is my focus." He admits, though, that the longer trips into the frozen wilderness, which can last up to forty nights, might no longer be possible. After all, scampering onto remote islands from a Zodiac boat to catch flap-happy penguins is probably best done when you're young. Still, when we speak, Neira was preparing a slate of new expeditions which would expand the scale of his work to multiple research stations across the polar continent. Beginning in the most remote place on Earth, he contends, is the best way to track the global trajectory of a virus that seems the likeliest to cause the next pandemic. Having witnessed H5N1's capacity to transform

Chile's coastline into a carrion landscape of rotting bodies, it is no wonder that Neira can no longer look away from the threats our species, and many others, face.

■ ■ ■

Among the chief imperatives of One Health is conservation. If we don't protect the natural spaces that wild animals inhabit, we'll create wave after wave of casual synanthropes entering our communities, which will amplify the number of transmission pathways that viruses can exploit to sicken and kill us. The Convention on Biological Diversity, a United Nations initiative, has even proposed a "30×30" goal for the globe: protect 30 percent of all wild spaces on land and on sea by 2030. It's an ambitious proposition, largely because of the repeated failure of political leaders to stand their ground in supporting conservation initiatives. If taken up at scale, though, it will no doubt ensure the survival of countless species in the short term while protecting humanity in the long run. The challenges are many, but if we are serious about preventing the next pandemic, the health and well-being of wild animals and their habitats must become a priority.

There's just one problem. How can you practice conservation when the creatures you're protecting want to eat you?

Part III

The Synanthropocene

Take a walk through your neighborhood and chances are you will cross paths with a synanthrope. Both of you, long cured of your neophobia, will either pay the other little notice or, more rarely, steady yourself for conflict. Once attuned to the wild animals in our midst, it's hard to unsee them: They are everywhere, populating every conceivable ecological niche that urban centers afford.

But the fact that synanthropes are present in cities doesn't signal that they are thriving. The age of urban centers is correlated with a lower density of animals, meaning the longer that cities are in place, the more likely it is they will become inhospitable to non-human life-forms. Cities, it seems, are like slow-moving ecological illnesses, replete with ecological traps that can doom resident species to extinction. But while they may tend toward dwindling biodiversity, that's not a sure thing. Cities are also uniquely mutable environments, capable of being radically transformed if humans set their minds to it. That means preserving species approaching the knife's edge of extinction must start in cities, because that's where so many of them will end up. It's also where we have the most power to remake the habitat to protect rather than destroy them. It is up to us, then, to imagine futures in which urban ecosystems are designed to help species proliferate instead of letting them die out, victims of passive city planning that fails to account for their needs or unwittingly sets traps.

Even when cities preserve species, they might not preserve their ecological functions. In the Dominican Republic, two species of globally threatened parrots (*Amazona ventralis* and *Psittacara chloropterus*) have flocked to cities, where they are present in population densities six times higher than in rural areas around the country—a clear sign they are thriving. But that's not the end of the story. The parrots play a critical role in dispersing seeds across the vast tropical forests of the Dominican Republic, but those that move into urban areas largely cease doing so. Instead of cracking seeds and spreading them across landscapes, which is a necessity for native plants to grow across the country, the parrots are content to eat trash and live a sedentary life. That adds another wrinkle to the challenge of transforming our cities into synanthrope-friendly spaces: Success isn't just ensuring that a high number of animals surround us but that those animals play ecological roles as well.

It's a tricky feat, but it's only the start of the challenges that conservationists face. Healthy urban ecosystems require animals to fill specific niches, like seed dispersal, pollination, or the removal of dead things by scavenging. We need synanthropes to serve as active participants as we build a world that celebrates complexity and interconnectedness. Those challenges are difficult enough when doing so poses no threat to us. They become infinitely more complex when conserving threatened species can cost us our lives.

Chapter 8

Embracing the Widow Makers

The hours are slow in passing as they always are
when you are waiting in fear for you know not what . . .

—Amitav Ghosh, *The Hungry Tide*

The Conservation Paradox. When wild animals and their habitats lie far away from human communities, conservation is a straightforward challenge and a self-evident good. In the Synanthropocene, urban areas are expanding into some of the natural world's most fragile and vulnerable ecosystems, forcing us into encounters with exotic animals, including endangered apex predators, that have been pushed to the brink. In these situations, conservation measures can amplify the threat animals pose to surrounding human communities. That's left some urban centers forced to make an unbearable choice: save human lives or doom charismatic megafauna to extinction.

■   ■   ■

Do you remember that leopard attack in the Indian courthouse? It was in the prologue, and we've shared a lot since then, so don't feel bad if it's slipped your mind. It was a terrifying moment, frozen on the internet, where it can be repeated over and over on demand. I've watched the clip dozens of times until the specifics of the situation have become as blurry as the footage. What stays with me is the sudden shift in the psyche of the humans involved, from hubris to abject terror. In the twenty-first century, technology is moving ever closer to magic, and animals, those quaint biological relics, should no longer be a danger to us. In fact, in any contest between humanity and other species, the ones most likely to die are those we've sent hurtling toward extinction. And yet, there was the Indian leopard, *Panthera pardus fusca,* rampaging through the courthouse, an attack that lasted four hours and ended up injuring at least five people as it tore its bloody path through the halls of justice. The human-modified environment of the courthouse, designed for our convenience, became a trap when the leopard adapted it to serve its purpose. The single corridor surrounding the building's square center gave the felid an unobstructed line to its prey, leaving the humans with no way to conceal themselves nor make an escape. But why had the leopard entered the courthouse in the first place? Where did it come from? And why was it so enraged?

The courthouse leopard wasn't a one-off. It was exemplary of a profound shift in big-cat behavior that's been happening across India. The attack happened in the Indian state of Uttar Pradesh, which is home to an astonishing 241 million people (3 percent of the world's entire human population) experiencing the highest population density of any state in India, with more than 2,500 people per square mile, about ten times that of the United Kingdom. Its population also is rising faster than anywhere else in India, and perhaps the entire world, with a turbo-charged growth rate of between 20 and 25 percent over the past two decades. The city of Ghaziabad, where the attack happened, is close to a wilderness area where leopards have long prowled.

Over the past few years, and just like their Carnivora cousin the coyote, Indian leopards have become entangled in a multitude of expanding urban matrices that have divided and subdivided their home ranges, with dire results: leopards found dead on the sides of highways and surface roads, killed after collisions; leopards attacking humans amid confusing urban fragments; leopards even eating babies. When I watch the video of the courtroom attack, I perceive the leopard to be in total control. The reality, according to terrified eyewitnesses, is that the precision of the violence it enacted belied its fear of being cornered within the courthouse's labyrinth of enclosed corridors. The attack wasn't the cool and dispassionate action of an apex predator. It was the terror of a hunted animal desperately fighting for survival with every weapon at its disposal.

Fear can make animals, including humans, do terrible, murderous things. And when there's fear on both sides, the results can spiral out of control. With an expanding population of weak but numerous humans facing down dwindling populations of powerful but increasingly rare apex felids, reaching a stable coexistence will undoubtedly require sacrifice. But how much are we willing to give along the way, and what value do we put on a human life when it stands in the way of saving a species from extinction?

▪ ▪ ▪

Though our species depends on healthy ecosystems to survive, it's only since the mid-1980s that the science of preserving them was formally created. The person most often credited for that is Michael Soulé, an eminent biologist who died in 2020 but whose long and distinguished career spanned the era of acid rain, unchecked industrial pollution, nascent economic globalization, and the arrival of climate change. In a seminal paper from 1985 titled "What Is Conservation Biology?" Soulé made the case that the study of ecosystems was a "crisis discipline," its relationship to biology akin to that of war studies to political science. The point, he argued, wasn't just that scientists should observe what was happening to the environment but that they had a duty to actively avert looming catastrophes. "In crisis disciplines," Soulé wrote, "one must act before knowing all the facts; crisis disciplines are thus a mixture of science and art, and their pursuit requires intuition as well as information."

Soulé's gambit was that the goodness of healthy ecosystems is so self-evident that it doesn't need to be proven scientifically. This led him to establish four "normative postulates" of his new discipline, a set of indisputable rules to guide it as it confronted the crisis of environmental destruction. The first postulate is that a higher diversity of organisms is good. The second is that ecological complexity is good. The third is that evolution is good. And the fourth is that biodiversity has intrinsic value. This last normative postulate, Soulé contended, is the most fundamental, because it emphasizes the inherent value of nonhuman life. "Species have value in themselves, a value neither conferred nor revocable, but springing from a species' long evolutionary heritage and potential or even from the mere fact of its existence." In other words, the deeply held belief that humans hold primacy over other animals is wrong. Instead, an ecosystem made up only of humans is less good than one in which other species participate—even when it means that fewer of us will survive.

It's hard to argue with Soulé's postulates. They are the ecological equivalent of Descartes's "*Cogito, ergo sum*" (I think, therefore I am): propositions so fundamental that they cannot be disproved. If you've ever visited a coral reef, you likely know what it's like to have your heart arrested by the extravagant beauty and evident goodness of nature. As someone who's been coral pilled, these habitats make me daven in religious obedience along with Soulé's script: *Biodiversity is good, ecological complexity is good, evolution is good, healthy ecosystems are inherently valuable.* Still, as a scientist myself, it strikes me as a strange way of crafting a scientific discipline, a task usually done by generating testable hypotheses and then seeing whether these hold up under rigorous scrutiny. Soulé's prescription for conservation science, which pulls from the artistic as much as from the rational, and is built on equal parts intuition and information, is diametrically opposed to the path of experimental science. There is no time to test the fundamentals, he argues, when action is so urgently required. And anyway, how would you even test them? It's like trying to test the value of the Ten Commandments. If we're to conserve Earth's wild places and inhabitants, Soulé implies, we must act now rather than quibble over self-evident truths. In an era of human-driven mass extinction, it is hard to argue with his single-mindedness.

■ ■ ■

You could find no better example of the self-evident goodness of non-human animals than the Bengal tiger, *Panthera tigris tigris*. Though only about four thousand still exist on Earth, the felids have an outsized influence on the human psyche. We know the tiger without ever having seen it: fierce eyes, formed of unbroken rings of yellow surrounding black pinpricks, a round head ending in a thick muzzle of ochre fur, and a Rorschach test of black stripes across its face, emanating immense psycho-spiritual power, creating a hypnotic array that draws prey in and freezes it in fear. Its gigantic muscular body is simultaneously lithe and stocky, its coat ranging from the off-white of wildfire ash across its belly and the insides of its legs and paws to a bright cape of Cheetos orange draped regally across its back, shoulders, and skull. Adult males are ten feet long and weigh more than five hundred pounds but can jump a distance of twenty feet in a single bound. And when the Bengal tiger jumps, mouth agape, its massive canines bared and ready to grip its prey, you understand instinctively that we live in the tiger's world: It is sovereign, all physical laws bent to its will, an eminence wielding supernatural forces that the rest of us can barely comprehend or come close to mimicking regardless of the technological magic we might muster.

Bengal tigers remind me that violence can be beautiful. It's a hard truth, as it punctures the story I tell myself that the most rarefied form of existence is nonviolence. The tiger's quick precision in taking down its prey, like the chital deer, an adorable brown oval on stilt-like legs, is a wonder to behold. We do not deserve this creature, this world. And perhaps nowhere is this more evident than in the Sundarbans, a river delta covered in mangrove forests spread across an ever-shifting archipelago of islands that straddle southeast India and Bangladesh. Here, within this region, human violence (blunt, cruel, and extravagant) and that enacted by the tiger (efficient, elegant, and necessary) have come face-to-face, with disastrous consequences for both species.

▪ ▪ ▪

It was a Tuesday in 2022 and Karuna Mandal started the day just like any other. It had been a hard few years. The pandemic had upended her community as a rush of people returned to the Sundarbans to seek their livelihood catching crabs and fish from the many tributaries of the Ganges, Brahmaputra, and Meghna rivers that converge here before flowing out into the Bay of Bengal. There had been more death than usual, with the sickness from COVID-19, and access to vaccines lagged across the region's poverty-stricken islands. And there were the cyclones, four in four years starting in 2019, cataclysmic events that destroyed mangrove stands and freshwater fishing ponds, inundated farming plots with salt water and rendered them infertile, tore apart dwellings, and destroyed the region's basic infrastructure for waste disposal, occasional electricity, and potable water. A little bit more land was lost each time the winds came as huge tracts of muddy coastline were torn away and submerged in the rivers, a ruthless and permanent theft. Bit by bit, the Sundarbans was being eaten away by the planet, never to be recovered. Throughout the cascading ordeals, though, Mandal remained steadfast in her devotion, starting every single morning by praying to Bonbibi, the Forest Mother and protector of this great outpost, for safe passage.

Bonbibi is an Islamic folk saint who has become the fulcrum of a syncretic local belief system here, drawing from Hinduism, Islam, Adivasi (the beliefs of India's Scheduled Tribes), and Christianity. Bonbibi is not interested in spiritual purity but in survival in the here and now across a landscape that demands respect and fear. Her origin story goes like this: Bonbibi, the daughter of a Sufi fakir from Mecca, is called upon by Allah to protect the "people of the eighteen tides," one of many names for the inhabitants of the Sundarbans, who are under threat by mystical forces. Their antagonist is Dokkin Rai, a Brahmin high-caste priest with the power to transform himself into a tiger. Bonbibi, whose name translates literally as "Woman of the Forest," is challenged by the Brahmin shape-shifter to a pitched battle, which she wins handily. As she stands victorious above him, Bonbibi chooses mercy rather than violence, offering to divide the Sundarbans in two: Half would go to humans, which she would protect from danger, and the other half would remain a wilderness ruled over by Dokkin Rai, who must pledge to never harm humans again as long as they keep their greed in check. The Brahmin agrees to her terms.

This is the story of the Sundarbans and its people, locked in an eternal duel with a murderous specter emanating from the natural world and protected by a mother figure who could destroy the tiger lord at any moment but chooses to spare him instead. It's a story less about victory over a villain than about a tenuous armistice that risks collapse daily because of our avarice. The uneasy truce will, of course, be broken by humans, who inevitably take more than their share from the mangrove forest and its waterways; this is just the way our species operates. In the meantime, the people of the Sundarbans must practice restraint, holding off that breach for as long as possible. It starts by praying to Bonbibi daily, regardless of religion or caste, while revering the Brahmin priest painted in orange and black stripes. And it means abiding by the rules of the accord: Human incursions into Dokkin Rai's domain must be kept to a minimum, and those that enter must only take what they need and nothing more. Transgressions will be punished by the tiger priest himself, who will eat you alive.

It's a beautiful origin story. For the millions like Karuna Mandal who rely on the mangrove forest for subsistence, it is also profoundly personal. Mandal has never entered the forest, but her husband, Dharani, scratched out a living there to support her and their two children. He caught fish and crabs from the tributaries flowing between the many islands of mangroves, bringing in an income that, while relatively meager, allowed happiness to seep into the couple's lives. His respect for the rules of the forest and its tiger lord meant that he would always see his wife again.

While her husband was off on fishing expeditions, some of which lasted days at a time, Mandal stuck to a steady routine. She's telling me this story through an interpreter while she sits on an old plastic chair in front of her home, a bricked-up square with windowless frames set in each wall. Short palms grow from the hard-packed sandy ground, and the area surrounding her shack is littered with rope, plastic containers, concrete bricks, and a broom handle. "I used to start my day at six in the morning," Mandal says in a slow and broken monotone, her words spilling out and then receding into silence like the tide, "and take our cattle to the pasture." After tying them up so they could graze on the sparse and salty grasses, Mandal would come back to her plot and mop the floors and walls of her home with cow dung, an ancient practice believed to keep mosquitoes, malaria, and bacteria at bay. "Then," she says, "I would cook for the family"—she and Dharani had a son and a daughter—"and then bathe in the pond near my house." At night, she explains, the electricity shuts down across the Sundarbans. "So I would turn on the kerosene lamp for a little light and stitch *kantha*," a local technique that repurposes old saris into multicolored quilts.

Mandal has spent her entire life in Chargheri, the most remote community on Satjelia Island, an inhabited island at the boundary of the Sundarbans National Park. A short walk from her home lies one of the many tributaries of the Ganges River; beyond it are the fifty or so forested islands protected by the Indian government, part of an archipelago of two hundred islands and at least four hundred tidal rivers, creeks, and canals, a collection of interconnected and mutable waterways subject to tides drawn from the Bay of Bengal that move across hundreds of kilometers each day, surging in all directions. The mangrove forest is a constant presence: animate counterbalanced provider, source of death and life. On the banks of the Garal River, the tributary nearest her home, the opaque and brackish waters lap a clay embankment made up of gray and glossy dunes lined with young mangrove saplings, a buttress against the howling cyclonic winds and the rising seas that continually threaten these islands. A rough concrete walkway runs circuitously from the dusty village down to the river's edge, where it ends in a couple brick steps used as a boat launch. The monochromatic scene—gray clay, gray water, gray mangrove roots growing aboveground, creating gnarled bell-shaped structures—is punctuated only by the vapid green of the sparse salt-soaked grass and the leathery elliptical leaves of the mangrove trees. From the shore of Satjelia Island, the river delta is a muted landscape: no towering trees, no drama of cresting hills and tumbling valleys. The murky water is a flat surface skidding over itself. The closest landmass, Marichjhapi Island, lies several hundred meters across the Garal River, and it provides few markers to situate you in space: more clay, short mangroves, and a low coastline at risk of being swallowed by the waves, just like the unlucky islands in this region that have been drowned already. The river and everything beyond is Dokkin Rai's domain—it is Dokkin Rai himself. Every time Dharani left his wife and ventured toward the improvised boat launch on his way to catch crabs, he put himself in the tiger lord's power. But Bonbibi would keep him safe so long as he kept his greed in check.

That day in 2022, with Dharani away catching crabs, Mandal took a morning bath in the nearby pond and came back home to do her chores. She began by praying to Bonbibi, for she knew that she had an important part to play in her husband's safety. While it is the man's duty to take only what he needs, it is the wife's to pray fervently enough that Bonbibi hears the calls for his protection. Like most people here, Mandal had memorized sections of the Bonbibi Johuranama, the holy book that tells the story of Bonbibi and Dokkin Rai, the sounds worn deeply into her psyche after decades spent surviving by the conditional benevolence of the forest. As she waited for her husband to come home, she began to chant the lines.

> *Bipod e poriya bon e jeijon e daak e,*
> *Ma boliya Bonbibi doya r maa take . . .*
>
> Facing any danger inside the forest, whoever prays to Her,
> Mother Bonbibi protects them all . . .

Suddenly there was yelling at the door, interrupting her recitations. "It was Dharani's two companions," she says, "but he was not with them." They were breathless and agitated, their story spilling out quickly. The three of them, they said, had been catching crabs in the mudflats of Marichjhapi Island along a tributary of the Garal River that led to a secluded part of the forest.

Midsentence, Mandal stops. I wait for her to continue. She sighs. She's wearing a bright orange sari with flecks of purple and a loose blue-black scarf draped around her neck. Her mouth is set in a lopsided grimace, and a deep frown cuts into her forehead. Her hands rest on her thighs, loose shiny brass bangles bright against the blue of the raised veins that run along her hands and the rough brown skin of her fingers, which hardly move. She sighs again before venturing on. "They were all in the *chandi,*" she says, referring to the gondola-like boats made of mangrove wood that fishermen use on their expeditions, "pulling up the crab traps out of the water." Fishermen here use many techniques to catch crabs, including casting nets from the shoreline, digging in the deep clay-like mud of the inner forest with thin and flexible steel rods to drag crabs out of their burrows, and—the technique Dharani used—throwing a series of linked and baited bell-shaped traps into the shallows near coastal mudflats along with half-brick sinkers. Each trap can handle only a few large crabs, each worth about thirty cents, while allowing smaller fish to escape, making it an efficient and targeted method in a region that prizes balance. Bonbibi provides.

Dharani and his companions had left their traps for a few hours and were heading back to learn how generous the forest had been. As they skimmed the river, they would have heard *bhatiyali,* hymns of devotion and protection stretching back centuries (the word translates as "to move downstream"), rippling across the water, or perhaps emanating from their own lips. These are local songs that can't be sung on dry land, needing rivers to bring them to life. In some cases, songs are sung by *boulays,* men with powers, both practical and supernatural, to keep tigers away. Their incantations, it is said, can control the movements of the felids, confine them in fixed circular spaces, or force their jaws shut. Sometimes, boatmen sing simply to create poetry from the water.

Amae bhashayili re, amaye dubayili re,

okul doriyar bujhi kul nai re /

Kul nai kinar nai, nai ko doriyar pari /

Tumi sabdhane chalaiyo re majhi amar bhanga tori re . . .

You have set me adrift,

you are drowning me /

This endless river without shores /

Steer the boat, cautiously o boatman,

this tattered boat of mine with broken rim . . .

The men on the *chandi* spotted a plastic water bottle, their improvised buoy, bobbing on the surface. They circled it and began the arduous process of pulling up the line of traps. But as they began to heave on the rope, something was wrong. The complicated series of slipknots that held the traps together had become tangled. No matter how hard they tried, the men couldn't pull them out of the water and onto the boat. Dharani, sixty-five at the time, had spent his whole long life learning how to catch crabs here. Without hesitating, he hopped into the water, feeling his legs sink into the familiar mud below, and patiently began untangling the ropes, just a few steps from where the flats rose above the water's surface, the clay fixed in place by young mangrove trees.

Things had changed in the decades since Dharani first learned to catch crabs from his parents. When he was a young man, the forest was its own barrier. The waters spilling in every direction erased footprints and telltale paths. The shoreline could transform itself within days as mangrove saplings shook off the gray patina of mud, rooted themselves in the deep clay, and unfurled bright new green leaves. Just as easily, acres could be destroyed overnight, washed away by heavy winds, extending the barren mudflats farther where before stood a healthy stand of trees. Long-standing tributaries leading a honey or wood collector into the heart of an island might evaporate in the chokingly hot summer months without a trace, though new waterways sometimes spilled open in the rainy season, a gift from Bonbibi or a trap laid by Dokkin Rai. The animistic magic of this place is so powerful that it can even swallow up whole islands. In 1996, Lohachara Island, once home to ten thousand people, was completely submerged by rising sea level; it's believed to be the first ever inhabited island destroyed by human-driven climate change, and its loss sent refugees scrambling to the mainland, where they were forced to start anew, their identities stripped away.

The landscape was still as changeable as ever when Dharani hopped out of the boat to free his livelihood from the rat's nest of rope, but the forest was no longer the only danger here. A thin skein of netted fencing ran along Marichjhapi Island, put there by forest guards a few years prior to stop Dharani and the hundreds of thousands like him who depend on the Sundarbans' natural resources from entering Dokkin Rai's kingdom, which had officially become off-limits in order to protect the region's tigers. Held aloft by long poles stuck into the mud of the shallows, the fence was hardly an impenetrable barrier, but it was a visible representation of the National Tiger Conservation Authority. Established in 2005, and better known as Project Tiger, the Conservation Authority had come to the region and immediately complicated the efforts of people here to keep themselves and their families alive. Along with fencing, the authority also established a network of armed forest guards who traveled by boat, patrolling the waters and detaining those who crossed into the forbidden territories of the tiger reserve, which had been officially established in 1973 but largely ignored by both government and locals until Project Tiger was launched. In doing so, the government had made the situation crystal clear. From now on, the state would serve up foot soldiers for the army of Dokkin Rai. In practical terms, that fencing meant places Dharani and others normally went to catch crabs were now off-limits, which forced them to seek out new and more obscure places to fish, often farther from home, which they could enter without the authorities finding them, but where other dangers lurked.

Chest deep in the water, Dharani steadied himself against the boat, his skilled hands working as quickly as possible to dislodge the knots. He was facing away from the nearby shore, his entire focus on freeing his catch. And then he was dead. Bengal tigers, as comfortable in the water as on land, are calculating hunters, stalking their prey from behind, never wasting an ounce of energy. The tiger had been watching from the shoreline unseen under the mangrove canopy, waiting for the ideal moment to attack. It saw Dharani facing away from shore in the water; it saw his companions in the boat. It leapt into the air, a blur of teeth and protruding claws, four hundred pounds of force slamming into his back at high speed. Dharani, his legs stuck in clay, had no way to escape. Within minutes the tiger had ended his life and robbed Karuna Mandal of her happiness.

From the deck of the narrow *chandi,* Dharani's companions attempted a last act of bravery in service of their friend. They grabbed the oars and swung them at the tiger, screaming at the monster as it tried to drag Dharani's body onto shore and into the obscurity of the forest. Dharani was a tall man, his legs still stuck, and the water deep enough that the tiger couldn't quite find purchase. The men in the *chandi* kept up their efforts, swinging the oars at the predator, screaming for it to release the body while the tiger resolutely tried to find a way to get its meal to shore. Eventually, harassed and exhausted, the beast gave up, rapidly swam the short distance to the shoreline, and disappeared into the forest, leaving the men free to collect their friend's corpse.

Karuna Mandal's story comes out in half clauses, with words suspended in the air, all of it in a pained blurry monotone. "It was a Tuesday," she adds blankly at the end.

Dharani's death was a tragedy and the first in a series of cascading losses that Karuna Mandal has endured. "He would stop at the market," Mandal says in a quiet voice, "and buy me bangles made of conch shells." And then: "He was such a good gardener." Later: "I think I was the happiest when Dharani and I played with our grandchild together. He was ten months old when Dharani died, and at least they had that time together, to allow Dharani to know that happiness." After a long pause. "I used to love to sit beside the river and play with my chickens. The air is so calm there, it would just relax me completely."

Since her husband's death, Mandal's sliver of happiness—a husband who cared for her and their children, moments of calm at the river's edge, the simple task of tending to five heads of cattle, the satisfaction of work done well, and the embrace of a community—has disappeared. In its wake, she has taken on an identity that has become legion among women here. Karuna Mandal has become a "tiger widow."

■ ■ ■

There is the story of Dokkin Rai's fury, of his lust for devouring flesh, and of his aristocratic domain over the living subjects of the Sundarbans, humans included, when they break the rules of good conduct. Or at least, that's one way to understand how the presence of tigers and humans in close proximity here have put members of both species at risk. There is another story to tell on that score: a 1979 massacre on Marichjhapi Island, directly across the Garal River from Karuna Mandal's home in Chargheri, of refugees who had fled neighboring Bangladesh when it declared itself an independent state a few years earlier, to escape persecution, mass killings, and sexual violence. After the partition of India and Pakistan in 1947, a radical left-wing political movement in West Bengal took up the cause of refugees from Bangladesh and supported their efforts to migrate into the Sundarbans. But when that same left-wing movement finally rose to power in West Bengal in 1977, things took a sudden turn. Instead of supporting the repatriation of Bangladeshi Bengalis, many of whom were Namasudras—an untouchable caste—the new state government abruptly reversed course and demanded that the Namasudras leave. Already, though, fifteen thousand refugee families had settled on Marichjhapi Island, transforming its dense forest into a thriving community with its own schools, roads, potteries, bakeries, fisheries, smithies, a health care center, a bustling market, and a working government. The only thing it lacked was drinking water, as the rivers coursing through this region are salty and brackish. By January 1979, the local government escalated its position, using thirty-one police ferries to create a total blockade of Marichjhapi Island, and killing anyone who tried to leave to find water. Refugee aid organizations quit the region and disease soon spread. Each time a boat departed the island for supplies, it was rammed and sunk by police, regardless of who was on board; women and children were frequently among the drowned. Thousands fled back to Bangladesh. Then, in May 1979, the state government launched a full-scale attack on Marichjhapi Island, setting fire to the community's market, hospital, schools, and homes, and murdering everyone they could. All in all, seventeen hundred people were killed over two nights of state-enacted violence. Thousands more had already died at the hands of police, from disease, or had drowned in the opaque and turbulent waters of the river delta. It is one of the gravest massacres in India's history and one that has been largely buried.

The state's extirpation of the Marichjhapi Island community was an act of merciless mass violence. It was also justified by the government of the day using a curious rationale: The Namasudra refugees were, the government stated, "in unauthorised occupation of Marichjhapi which is part of the Sundarbans Government Reserve Forest violating thereby the Forest Act." Specifically, the government claimed that the island was critical to a World Wildlife Fund–sponsored tiger conservation project, set up to manage the charismatic megafauna's critical habitat. It was a horrific perversion of Soulé's work: mass slaughter justified by the normative postulates of conservation science. *Biodiversity is good, ecological complexity is good, evolution is good, ecosystems are inherently valuable.* These self-evident truths, it would seem, made undesirable humans expendable.

There is a macabre coda to the Marichjhapi massacre. According to eyewitness accounts, when the killing was done, the government piled up bodies of the dead and injured onto boats. They then brought them into the heart of the tiger reserve and dumped them there for the animals to feed on. This is, according to local stories, when the felids developed an appetite for human flesh.

The myth of Bonbibi and the Marichjhapi massacre, when refugees became "tiger food," aren't the only ways to explain the growing legion of tiger widows like Karuna Mandal. There are other stories here, as numerous as the region's rivers, creeks, and canals, which also merge, intersect, shift, and evolve with each passing day. Bengal tigers, goes one, have an appetite for human flesh because of the region's brackish water; its saltiness destroys their livers and kidneys, drives them half mad, and makes them thirsty for blood. Another says that the long-standing Hindu custom of cremating the dead upstream on the Ganges River, whose tributaries run through the Sundarbans on their way to the Bay of Bengal, ended up spilling human remains into tiger territory, habituating them to our taste. When the Farakka Barrage, a massive dam, was built in 1970, the tides no longer deposited burned corpses at the tigers' feet, and the felids began to hunt live prey instead. Still another claims that poaching of native chital deer and wild boar in the Sundarbans forced the Bengal tiger to look for other large and defenseless animals to prey upon. Humans, stuck thigh deep in sucking mud, hands tangled in ropes, were a perfect substitute.

I'll venture one more. The Sundarbans might seem like a far-off exotic locale, remote and disconnected from global patterns of accelerating urban sprawl. And while this place has vast swaths of protected land, the western edge of the four-thousand-square-mile reserve is bordered by villages and towns that are rapidly becoming part of Kolkata, a city of 15 million people that is expanding toward them at a blistering pace, transforming the 4.5 million people who live here into citizens at the fringes of one of the world's largest metropolises. The pressure of Kolkata's urban sprawl would naturally displace the local population here farther out across the archipelago of islands. But it cannot move: The tiger reserve, encompassing the islands and rivers east and south of Chargheri, is off-limits to development (and the punishment of unauthorized settlement is still, five decades later, a living memory here). Since 1991, the Sundarbans has also lost about fifteen square kilometers of vegetation each year as people, always resourceful, turn natural habitat into urban space. Meanwhile, the delta spills out into the Bay of Bengal, but the bay itself is also claiming more land, with about two hundred meters of coastline also lost annually from erosion and inundation.

That has turned the boundaries of the tiger reserve into something like the U.S. 101 freeway in Los Angeles, which impedes mountain lions from expanding their limited home range and evading extinction.

Here the roles are shared between *Homo sapiens* and tigers. The people of the Sundarbans are squeezed by intensifying urban pressure on one side and hemmed in by rivers on the other. Worse, the islands they live on are eroding with every cyclone and storm, leaving less space for rapidly urbanizing communities while ratcheting up the pressure for residents to enter the forest for fish, crabs, honey, or wood to feed their families. On the other side of the river, the Bengal tigers are scattered across dozens of isolated islands in minuscule population units surrounded by bodies of water often too large for them to ford (they can't swim more than about a kilometer at a time). This leaves outposts of tiger alleles that can't be exchanged, hastening an extinction debt that might spell their end in a few generations' time. There is no hope for tigers that leave the forest reserve in search of new genetic material, because we surround them. To succeed, they would have to navigate through millions of humans packed tightly together to link up with other tiger populations spread across southeast India. Still, every year or so, a tiger tries its luck and invariably ends up facing down hundreds of humans armed with sticks, bricks, machetes, heavy rope nets, and a blunt mob intelligence. It is a sobering sight: a regal beast brought down by an organism that lacks its power and commanding beauty but slowly weakens it through base ubiquity and an unrelenting bloodlust.

All is not lost. Not yet. Though slightly more than two hundred tigers roam the thousands of kilometers of the Sundarbans reserve, their population has a surprisingly high level of genetic diversity. This is measured through a test called "expected heterozygosity," which refers to the likelihood that any one individual will have a different set of genes than any other. Among Bengal tigers in the Sundarbans, this ranges from 58 to 72 percent, meaning that at minimum there's a 58 percent chance that any one tiger in the region will be genetically different from any other. Compare that to the mountain lions in Los Angeles trapped south of the U.S. 101, whose expected heterozygosity has dipped to 29 percent, meaning that more than two-thirds of the population are effectively identical—a death knell for the population, barring a dramatic turnaround. Based on their genetic diversity and the blessing of protected mangrove forests within which to roam, the population of Bengal tigers in the Sundarbans looks better equipped to evade extinction than other threatened apex felids. But in a region in which you can't even count on rivers to follow the same path from one day to the next, nothing is truly certain.

▪ ▪ ▪

Desperate people and desperate animals are a recipe for conflict—one that has been well studied by urban ecologists. Animal attacks on humans fall into three categories: predatory, territorial, or defensive. In urban centers, predatory attacks are rare, but they do happen. In most cases, it's large carnivores that end up attacking humans, and scientists have identified a disturbing linear relationship that connects the two: Animals that more frequently predate on humans are also more likely to be close to extinction. It's a dangerous feedback loop. Large carnivores, starved because of climate change and habitat loss, are pushed into conflict with humans, thereby further hastening their decline.

Attacks by large carnivores happen all over the world. While rare, they are not random. In high-income countries, humans are overwhelmingly at risk when they're doing recreational activities like hiking, running, or camping. In poorer places, the attacks almost always happen while people are doing things necessary to keep themselves and their families alive and fed: farming, grazing livestock, fishing, and collecting materials from forests. It's a bit like the luxury effect that Chris Schell described, in which urban biodiversity increases in wealthier neighborhoods, except this linear relationship stretches across global regions instead of cityscapes, and brings terror instead of the chance to commune peacefully with nature. Follow that trend to its most extreme outlier and you end up in the Sundarbans, one of the poorest places on Earth, where animal attacks are as predictable as the shifting tides. This is the story of Dharani Mandal, dead because of a tangled rope. And it's the story of so many other men claimed by tigers, a legion of widows left in their wake.

Along with Karuna Mandal, I spoke with many other tiger widows who lived at the teetering edge of destitution until a tiger attack shattered the last remnants of stability and left them reeling. Satibala, clothed in a green and purple sari, her face a perpetual scowl, also lives in Chargheri, in one of the village's many corrugated tin-roofed shacks; she sits barefoot on red bricks on an improvised stoop surrounded by hardened, salt-choked dirt. One day her husband, Santosh, was casting nets across the shallows of the Harikhali mudflats, deep in the tiger reserve near the border with Bangladesh. It's a common strategy used to pick up smelt, small crabs, and larger opportunistic fish that populate the intertidal zone, but it can be risky. The shallows are most abundant at dusk, and Santosh had waited for the sun to set to maximize his catch. A tiger attacked him from behind, killing him and dragging his body, which was never recovered, into the mangrove forest. Since then, Satibala has been forced to take his place, catching fish and crabs in nearby mudflats. Without the financial support of her husband, she had to take her young son and daughter out of school. They found jobs in the heart of Kolkata, less than fifty miles away but requiring a seven-hour voyage; she says they are working as laborers, but the truth may be more tragic than she lets on.

Namita still hasn't replaced the vermillion dot on her forehead with a black one, as is the tradition for widows. There's a string knotted above her left elbow—a fabric worn by Hindus as a symbol of deep connection with a loved one. Namita's husband, Santu, stopped fishing with other people during the COVID-19 pandemic: There were too many new faces that had migrated back to the region after the labor market shut down in Kolkata, and Santu was worried both about catching the virus and navigating dangerous terrain with people who weren't attuned to its many moods. Like most fishermen, Santu didn't have a boat license certificate, which meant that he didn't have the right to fish inside the Sundarbans reserve. To elude the forest guards, he would fish late in the day when there were fewer patrols. One Sunday, he had gone out alone as usual, planning to be back to sell his catch at the *haat,* the weekly night market. He never returned. Namita, concerned for his safety, sent his nephew out to look for him. All that he found was an empty boat drifting near Marichjhapi Island. In the crisis state that followed, Namita was unable to care for her three young daughters and married them off rather than continue their education.

These are not isolated incidents. They are part of a trend, stretching back a century, of a growing legion of tiger widows, who now number thousands strong. According to one study, most tiger widows are in their early forties, but fewer than 5 percent of them are literate, leaving them with few viable income sources when their husbands die. Most live far below the poverty line and in their desperation will often take the place of their late husbands on harvesting expeditions into the mangrove forest, leaving them at risk of wildlife attacks and vulnerable to shocks, like the cyclones that tear up this place.

The precarity that tiger widows face isn't only economic. I met a local geographer who described widows as "inauspicious beings," a designation that sets them at the bottom of a social hierarchy still very much governed by a caste system in which social roles are immutable and highly prescribed. In the Sundarbans, a widow is often blamed for her husband's death, fueled by suspicion that she did not pray sufficiently hard enough to Bonbibi for protection. Shattered with grief and left destitute, a widow also encounters a community that often shuts her out of social and cultural activities, forcing her to fend for herself. Her friends stop talking to her. She's forbidden from participating in *puja,* festivals of worship that are central features of life and connection. Even seeing the face of a widow here portends bad luck, so family members often ask the woman to never leave her home.

In the rare instance when a widow ventures out, she is expected to cover her face to protect others from the evil and shame that she emanates. "They are," the local geographer told me, "so humiliated." This isolation has created deep-seated anxiety among tiger widows, who score over twice as high on a scale measuring the stigma they experience compared to widows in the region who have lost their husbands to snakebites or natural causes. No wonder, then, that half of them suffer from major depressive disorder, persistent depressive disorder, chronic pain, and post-traumatic stress disorder. Ultimately, tiger widows are left grappling with a dual crisis: finding a way to survive alone while searching for a reason not to die.

"If you go to any fisher in the Sundarbans region, they will say that the attacks are intensifying. Why? Because of the success of the conservation efforts, the number of tigers has increased over time." Prama Mukhopadhyay is an environmental sociologist and Kolkata native who spent years in the Sundarbans studying the ways that tiger widows and their broader communities navigate the shifting landscape of aggressive felids, state intervention, climate change, social exclusion, and poverty. "But," she says, "the number of people going inside the forest has also increased. So if the number of tigers has increased and the number of people going inside the forest has also increased, then the number of human–animal conflicts increases. That is something that logically we can understand."

Mukhopadhyay visited one small village called Bandhob-pur, among which 133 households relied on fishing for survival. Of those, almost 80 percent were home to a tiger widow. Mukhopadhyay undertook a years-long ethnographic study of tiger widows, which included embedding herself for months on end with women and families living in the region. "The Sundarbans is always made to feel like this exotic land out in some other place, far away, but it isn't." Instead, it's just one more urban appendage of Kolkata. "The city has been growing, and the Sundarbans is nothing but the city now—it's just on the fringes, not too far." Mukhopadhyay grew up in a neighborhood that was, she says, a part of the Sundarbans mangrove forest until about a hundred years ago. Nowadays, it takes only two hours to get from her neighborhood to the Sundarbans by train, and she spends the entire journey watching dense urban areas and sprawling slums pass by her window.

Kolkata, like all cities, is an early successional habitat frozen in time. Its reshaping of the environment is so total that you'd be forgiven for believing that there was never a mangrove forest here, only tenements and cracked concrete highways that will last forever. But unlike most cities, there's something dangerous lurking right across the river from Kolkata's farthest fringes. "It's very interesting to me that in our country," says Mukhopadhyay, "which is supposed to become one of the world's global leaders, that very nearby this urban center is a place where such things happen." "Such things" is the murky term Mukhopadhyay uses in casual conversation to describe the phenomenon of human–tiger conflict here. When it was time for her to process what she witnessed, she used a more direct descriptor: "everyday death." "I was very moved with the idea of how death is just there," she says, "without having to really talk about it too much—it was just really very everyday and mundane."

If your government's goal is to propagate man-eating tigers next to an urban area home to 4.5 million people, you're dealing with a certain inevitability. It's not whether deadly tiger attacks can be prevented but what can be done to minimize the social unrest that will foment when they occur. If the disruptions become too frequent, they might derail the entirety of the state-run Project Tiger, lead to more mob violence against felids, and ultimately end with the extinction of the species. The stakes are high: A singularly arresting apex predator forged over billions of years of evolution might be obliterated on the only planet we are sure harbors life. The death of the Bengal tiger endling—the last remaining member of a species—through the unchecked human compunction for progress would set the karmic clock reeling. In the Sundarbans, though, when a tiger and a person meet in space and time, it almost always spells the end of the human life. Nature is still terrifying; no wonder that our species has spent the past thirty thousand years or more feverishly trying to decouple our fate from that of the natural world. And yet, I can hear Soulé's voice tut-tutting me to remember his normative postulates, ideas so self-evidently true that they can brook no argument: *Biodiversity is good, ecological complexity is good, evolution is good, ecosystems are inherently valuable.* No amount of human death will prove them false. The Indian government evidently agrees.

It's not just that people die frequently in the Sundarbans, though they do. It's that death in this half-wild, half-urban landscape mimics its rivers and tides: It splits, reshapes, and unmoors, reconfiguring the orderly existence of those left behind into something that moves with an entirely different gait. A death can instantly transform a tiger widow's life from relative stability to total social isolation and financial ruin; in Karuna Mandal's case, the cascading shocks have also left her partially paralyzed. As tragic as that is, the death itself is only the beginning of the story. That's because proving that a man died at the hands of a tiger has profound implications for those he leaves behind.

The average annual income in the Sundarbans is less than eight hundred U.S. dollars. Most people survive on between one and two dollars a day, made primarily through fishing, catching crabs, collecting wild honey (done by smoking bees out of hives located deep in the forest), or harvesting hard and durable mangrove wood—activities that necessarily lead people into the home range of Bengal tigers. The people of the Sundarbans understand this and so does the Indian government. Both sides also know that there is no way to cease people's reliance on the forest for survival. The next best thing, then, is to take the supernatural problem of death and transmogrify it into one that the government can control: the transfer of money. When a person is killed by a tiger in the Sundarbans forest reserve, the West Bengal Forest Department provides their family with 500,000 rupees (about US$6,000), equivalent to about eight times the annual median income. It's a massive amount of money relative to this place and in principle should set a tiger widow up for a life of comfort amid their grief. It's a kind of alchemy: a corpse turned into gold. Like alchemy, though, achieving it in practice has proven elusive.

Since the compensation system was put in place, tiger widows have been denied their claims over and over again. The trouble stems from the boat license certificates that all people entering the forest must legally carry. The number of these certificates—known locally as "resties"—has remained static since colonial times, though the number of forest workers placing themselves under Bonbibi's protection has ballooned. Exact numbers are hard to come by, though there appear to be more then 150,000 fishermen who actively harvest in the tiger reserve, who collectively support the livelihoods of about 2 million people. But there are only about 700 resties, meaning only 0.5 percent of the total number have entered the forest legally on any given day. Purchasing one of the rare available licenses is prohibitively expensive, as is renting one from a resty owner, making practically the entire class of fishermen illegal workers subject to penalties and harassment by state agents. They can be stopped by a patrol boat, fined, and shaken down or worse during the hours and days they spend in the tiger reserve, rendering them even more destitute than before. And if they are mauled and eaten while they are working, the fact that they were there illegally will prevent their widows from being compensated. What's more, if someone is known to have died in the reserve without a resty, their widow can be held liable for their death and pay hefty fines. That makes widows unwilling to officially admit that their husbands are dead. "It is so difficult," says Mukhopadhyay, "to even come to terms with the fact that some people die with no record of that death, because if there is a record, that is a problem." In a country in which death is highly ritualized, tiger widows are stuck in ambiguity, officially denying the grim tragedies that have destroyed their lives and never able to truly grieve or let go.

When Dokkin Rai, the tiger lord, set a human life as the punishment for greed in the mangrove forest, he surely never anticipated how far-reaching the consequences would be. Often family units will break down after a tiger has killed a man. Without the meager but steady income produced through illegal fishing, the children of tiger widows confront a perilous future in which they must earn their keep to survive. There are an estimated 450,000 female sex workers in West Bengal, about 25 percent of the total number that work in India, despite the fact that the state makes up only 7 percent of the country's population. It is poverty that forces them into the trade and living in Sonagachi (Tree of Gold) and other red-light districts across Kolkata. Along with hurricanes, tiger attacks act as catalysts that funnel women and their daughters into the sex trade, severing the last remaining protection poor families have from utter destitution. In a recent report, India's National Crime Records Bureau noted that 44 percent of all human trafficking cases in India take place in West Bengal; one-third of the victims are women and girls from the Sundarbans.

The few hundred Bengal tigers in the Sundarbans have been given a fertile ten-thousand-square-kilometer area to keep themselves alive. It's a huge tract of land teeming with natural resources that could otherwise be extracted to lift people out of poverty and save the thousand or so women who become trapped each year in Sonagachi and other dark urban places. The conservation measures enacted by the Indian government meant to protect tigers have made the lives of the people who live side by side with the animals even more desperate. And still, I can hear Soulé's normative postulates whispered in my ear. *Biodiversity is good. Ecological complexity is good. Evolution is good. Ecosystems are inherently valuable.* These are truths beyond argument, or so we are meant to believe. But as I listen back to my interview with Karuna Mandal and the other women who spoke with me, their voices cracked and affectless, their husbands dead and their daughters gone, my faith in the self-evident goodness of these postulates has been shaken.

Chapter 9

Listening to Animals

By its nature, the metropolis provides what otherwise
could be given only by traveling; namely, the strange.

—Jane Jacobs, *The Death and Life of Great American Cities*

Synanthropes Expand Human Consciousness. Cities are designed
as static places, with the false hope that they can be oases of human
control. The success synanthropes have had in finding ecological
niches in our communities reveals just how poorly cities function to
keep animals out, while also failing to provide those that live among us
with sustainable roles in urban ecosystems. As we reach a breaking point
in our relationship with the planet, we need to recast cities as nodes
of biodiversity, adaptation, and species preservation rather than continue
the myth that they are biological deserts—static environments that lack
the necessary ingredients to sustain life. Synanthropes reveal different
ways of living in cities. Listening to them will show us how to redesign
human spaces so that animals—including *Homo sapiens*—flourish.
What we hear will guide us toward a more harmonious future as
well as a return to the deep intimacy that ties our species to
every other life-form on Earth.

■　■　■

Cities are massively disruptive phenomena, and at their worst, they're accelerating the decline of innumerable species across the world. Simultaneously, they're highly biodiverse places where species are congregating in ever greater numbers, and they're spurring adaptation and evolution at a blistering speed. We can look out at our cities and see them as annihilating the planet. Or we can move through our communities attuned to the wonder—the real live nature—that is flourishing in our midst.

Like a lot of people, I instinctively bemoan the changes I see happening around me: gentrifying neighborhoods losing their edge, more land claimed to an unwavering urban grid, and featureless new builds as anodyne as mud. Look up, though: There's a squirrel leaping, for no discernible reason, between two impossibly slim tree branches, jacked up on epinephrine and risking its life for the hell of it. There are urban foxes rutting in the garden a few doors down, screaming ecstatically, as loudly as they possibly can in the middle of the night (can you blame them?). Coyotes are hunting rabbits, sending slow shoots of soft black fur into the air. Raccoons are spreading our refuse—which we tastefully try to hide behind black plastic—for all to see. It's annoying. Sometimes it's frightening. And in some cases it's downright dangerous. But life is chaos and there is no going back. Building a denial of these basic facts into our urban design will make us less able to contemplate and come to peace with our own inevitable end and the constant transformation of the planet around us. Embracing change can sometimes feel impossible, especially when the changes all around us seem to be steadily making the world a more frightening and joyless place. But being with animals, wherever they find us, is a good starting point to commune with the parts of the world we've forgotten we love.

I've mentioned before that cities are early successional environments, an ecological term for the habitats that temporarily emerge after mass disruptions like forest fires or avalanches. Except, of course, they're frozen in time. We've designed them to be as static and predictable as possible to reduce the amount of danger and stress we face each day. One of their primary functions in that regard has been to keep other animals out. It's a sound strategy to preserve and isolate our species, but as urban sprawl has continued to delete and replace wild spaces across the world, cities have utterly failed to keep other species away. Instead, urban spaces are ensnaring a vast cross section of the world's wild animals, robbing them of their ecological functions, and sometimes even sickening and killing them in ecological traps. But that doesn't mean they have to stay that way.

The many synanthropes I've written about in this book along with all the others that live among us share a common thread: They have transformed urban environments designed to be hostile toward them into places in which they can survive. That they have done so is both a miracle and a reminder of the limits of our species in refashioning the world in our image. Imagine now that our cities weren't designed to keep animals out but to sustain them as healthy and important parts of the urban human experience. What would our days be like? How would our landscapes be changed? And what connections would we feel as we moved through the world? There is a future available to us in which we can normalize our relationship with synanthropic animals, be they raccoon gazes, neophobic coyotes, or even hordes of baboons. The first step is to recognize our close intimacy with a multitude of species and to treat their homes—our homes—as ecosystems worth saving.

Ironically, while the built environment of urban centers is static the world over, cities and towns are where the action is. We've grown accustomed to human unpredictability and have created in cities places that can accommodate people being people—eccentric, absurd, neurodivergent, social, introverted, anarchic, aggressive, kinky, devout, striving, and all the other ways of being that expand our collective consciousness and propel human civilization toward an unknown future. We've learned to revel in differences and bridge them when the distance is too pronounced. No creature has generated more chaos (and cruelty) than *Homo sapiens*. If we can live side by side, creating complex human social ecologies in the process, surely we can find a way to do so with other species. The next step is to transform our cities so that we can embrace the wonder and spontaneity of interspecies connectedness and share our wildness with others.

But let me tell it another way.

■ ■ ■

One day, during the mythic prehistory of the ancient Greeks, Melampus of Pylos, a famous seer, was called to the king of Thessaly. Iphiclus, the king's son, was infertile and the king feared that he would never bear children, thus dooming the king's dynasty to oblivion. Melampus considered the problem, which wasn't the first that a stressed-out sovereign had begged him to solve. He had a well-earned reputation for dealing with metaphysical conundrums and even predicting the future of the *oikouméné,* the known world, all of which stemmed from a singular gift: When animals spoke, Melampus understood them. It was a power he was given when two snakes licked his ears clean in gratitude after he spared their lives from the bloodlust of his irate servants.

Melampus, always a curious fellow, decided to help the king. He sacrificed a bull to Zeus and sat waiting next to the charred corpse. Soon enough, two vultures swooped down and gossiped while they poked at the flesh and bones (my guess is they were casual synanthropes, splitting their time between the urban matrix of the Thessalian city-state and the contiguity of the vast forests that lay beyond). Eventually, the birds' conversation fell to the topic of Iphiclus's delicate condition. It all began many years ago, one of the vultures explained to the other, when Iphiclus's father had been castrating goats. At one point, the king had raised his bloody knife in the air and the boy Iphiclus, catching sight of it, had panicked. In a frenzy, the Thessalian king thrust the bloody knife into a wild pear tree and quickly fled the scene with his distraught young son. The problem, the vulture explained, was that the king had inadvertently stabbed the knife into a hamadryad, a wood nymph (and the source of the hamadryas baboon's name), who was living in the tree. In revenge, the hamadryad cursed the boy with erectile dysfunction. Melampus listened intently to the scavengers' chatter. When they were done, he got up, tracked down the tree, extracted the bark-covered knife, and eased the hamadryad's suffering. Melampus then made a potion with the rusty blade and forced the overwrought Iphiclus to drink it. Soon after, the prince bore his happy wife two sons. The king was ecstatic. Melampus was richly rewarded.

Myths, like cities and synanthropes, are designed to evolve. They find new resonance each time this moment right now tips into the future, the stories serving as archetypes we can use to navigate new challenges that are often just ancient ones cast with new patinas. In the midst of the Synanthropocene, the myth of Melampus strikes me as a user guide to making cities healthy, balanced, and welcoming places for synanthropes, which in turn will make them more dynamic and beautiful places to live. In my reading, Iphiclus is the embodiment of the city, his infertility mirroring that of our urban places: drab and concrete early successional environments that have the power to flourish but, at their worst, will end up extinguishing the generational dynasty of living creatures. The king's knife stabbing the hamadryad is humanity's original sin: that we have built our communities with a total disregard for the natural habitats they've replaced. What the myth tells me is that unless we find the knife and heal the wound it's caused, we will never create a fertile, biodiverse, and happy future on this planet. The way there is to follow Melampus and attune ourselves to the language of animals and learn from them how to render our cities into places pulsing with life of all kinds.

Synanthropes have shown how even the most unlikely creatures can rapidly adapt to seemingly impossible urban challenges. If raccoons can create a city within a city without opposable thumbs and only a second-rate carnivore brain, surely we can transform cities into places where species flourish rather than go extinct. The challenges that synanthropes pose to us—animal overpopulation, predator attacks, and the risk of viral spillover events—call for creative solutions. But underpinning all of them is a shift in our own thinking. So, before we get to solving the problems that synanthropes cause us, let's remove the knife and correct the harms we do them.

Collisions are the number one reason that synanthropes die. Retrofitting our communities so that we don't absentmindedly run them over would go far in normalizing interspecies relations. Some studies have found that more than half of urban raccoons are killed on streets and highways. The numbers are even more dire for synanthropic megafauna like black bears (a fixture at dumps and garbage bins in the Canadian Northwest and the U.S. Northeast), as well as coyotes, bobcats, and white-tailed deer living in and around urban areas: Two-thirds of them die after colliding with cars. It's a hard problem but not a complicated one. We just need to follow the animals. Raccoons and coyotes, probing urban fragments, regularly repurpose culverts into shortcuts between fragmented habitats to avoid the killing fields of fast-moving freeways. It's a reminder that animals often can't survive by moving along city grids like us, and that corridors, even modest ones, can help enormously in keeping them alive.

▪ ▪ ▪

Saving urban animals from senseless deaths is a vexing challenge but it pales in complexity to stopping them from senselessly reproducing. Just ask a New Yorker. In 2024, after decades of losing a war on rats, New York City's disgraced then-mayor Eric Adams hosted the United States' first ever National Urban Rat Summit, an attempt at bringing some of the world's smartest rat people together to stop the ubiquitous rodents from messing with the Big Apple. It's a tale as old as time—older, actually, if by time you're referring to the moment our species began to measure its passing. If you remember the prologue, there's evidence of synanthropes (ravens and foxes, specifically) creating adaptive and sustained niches in human-modified environments 27,000 years ago, when we could barely congregate in groups larger than a couple dozen. Even as far back as 8,000 years ago, *Homo sapiens* had already lost the war on rats before it had even really begun. Since then, the victories of *Rattus rattus* have been mounting with no end in sight. Until the National Urban Rat Summit.

On the first day of the summit, on September 18, 2024, one presenter wasted no time in setting the record straight. In a presentation titled "Why Your 'War' Won't Succeed," Canadian ecologist Dr. Chelsea Himsworth laid out the case for abandoning victory in favor of a managed retreat. Killing rats with rodenticides, dry ice, euthanasia, electric shocks, snap traps, or whatever other macabre inventions might come along won't ever turn the tide, she argued. At most, it'll just keep rat numbers momentarily at bay. That's because rats, Himsworth explained to her rapt audience, aren't the real problem. Instead, as one of her slides read, "Focus on interfaces not entities." In other words, what everyone has thought of as a "rat problem" is in fact the thoroughly human problem of how we organize urban spaces. You can kill thousands of rats, but if the urban habitat remains unchanged, more will just keep flooding the zone. According to Himsworth, we will only meaningfully reduce rat populations if we shift our thinking toward urban relationships rather than single animals. Simply put, it's not the rat we must target but the rattiness we bring upon ourselves.

Himsworth studies rats in Vancouver's Downtown Eastside, a neighborhood that has long faced widespread homelessness, poverty, and public drug use. In the 1990s, after decades of failures in improving conditions in the neighborhood, city officials turned to a strategy of "harm reduction," an approach that recognizes the inevitability of ongoing drug use for some people and consequently shifts focus to reducing the harms it might cause. Himsworth has extended that philosophy to *Rattus rattus*. Trying to kill rats in decrepit tenement housing stock owned by slumlords is as futile as the War on Drugs: The rats, like the drugs, will always find a way into the heart of urban areas. To win the day, Himsworth contends, cities must target the housing itself by creating bylaws that enshrine the right of tenants to sanitary living conditions. That forces authorities to act—not directly on rats but on the long-standing neglect that has allowed the adaptable rodents to take over dodgy buildings and threaten the health and self-respect of the people who live there. And by targeting property owners, municipalities can force real and sustained change across entire neighborhoods rather than trying to clean up individual apartments.

To hear Eric Adams tell it, New York City's renewed War on Rats was the exact opposite of what Himsworth is advocating for. In announcing his intention to hire a "rat czar" to clear the city's streets, Adams posted a statement on X that made it clear where he stood: *There's NOTHING I hate more than rats. If you have the drive, determination, and killer instinct needed to fight New York City's relentless rat population—then your dream job awaits.*

The fighting words, it turns out, were just a smoke screen. Under Adams, the city's anti-rat strategy wasn't predicated on killing as many rats as possible. Instead, it followed the harm-reduction framework that Himsworth laid out: undoing the human-created conditions that allow the rodents to proliferate. The key innovation was "containerization," a five-dollar word for forcing single-family homes and small apartment buildings to put their garbage in locked bins instead of throwing bags out on the street where rats can easily devour their contents. It's a simple approach that dispenses with rats completely in favor of changing the urban environment itself. After the rules took effect in November 2024, rat sightings dropped by about 25 percent within four months. In a ten-block section of West Harlem, where a pilot project lined sidewalks with massive gray bins about half the size of a sedan to containerize every iota of trash the neighborhood produced, results were even more dramatic: a 60 percent drop in rat sightings within a month.

The Sarawat Mountains of Saudi Arabia, where drug dealing is punishable by death, could not be more different from Vancouver's Downtown Eastside or West Harlem. Still, Ghanem al-Ghamdi, the veterinarian, and others here have adopted what amounts to a harm-reduction approach to controlling the marauding baboon hordes that threaten to overrun the mountain city of al-Baha and the pilgrim-thronged city of Mecca. This isn't the first attempt. In the 1980s and 1990s, al-Ghamdi says, the government instituted a plan to shift the behaviors of baboons, which included poisoning and gunshots. It failed miserably. Funding was scant, and the collective human will to stop their incursions wasn't there. "This time," he says, "the government is trying to do a better job." It's also on the clock: In February 2023, the kingdom committed itself to "solving the baboon problem" by 2026, one of many ambitious projects Saudi Arabia is undertaking as part of Vision 2030. (Other projects include The Line, a 170-kilometer-long and 200-meter-wide tunnel-like futuristic "smart city" to be housed in a single narrow building dug out of the desert, and a truly audacious attempt at sports-washing its atrocious human rights record by campaigning for LGBTQ+ visitors during the 2034 FIFA World Cup, despite the country sometimes punishing same-sex acts by death.) Just like Adams-era New York City, the new project targets the relationships between baboons and humans, through three steps: restoring ecological balance between human-modified and pristine habitats, curbing human behaviors that lead to baboon invasions and attacks, and gathering detailed data on baboon population dynamics. Al-Baha is a key testing ground for the new plan, and al-Ghamdi is playing a central role by mapping baboon hot spots and coaxing human behavior change.

When I visited Raghadan Forest Park in al-Baha, a line of over-turned blue garbage bins stretched to the horizon, the handiwork of about twenty-five baboons that rampaged around me. I was amazed at the total freedom the baboon clan enjoyed: The many staff tending to the park that day were completely unmoved and didn't bother to inter-rupt their general mayhem. Still, al-Ghamdi was adamant that his efforts were working. "The numbers," he said, "are a little bit decreasing." He wasn't being Pollyannaish either: When his team had run a population census three years earlier, Raghadan park was seeing hundreds of baboons daily rather than the couple dozen with which we had crossed paths. The differences were even starker at a nearby supermarket at the edge of a cliff a few hundred yards from the park. Massive, uncovered dumpsters teeming with the market's waste had become a source of food so reliable that hundreds of baboons had moved from a sleeping site outside of al-Baha to the sheer cliff face below the supermarket. In the old days, al-Ghamdi said, the government would have just shot them, a short-term fix that would have done nothing to solve the underlying issues driving the relationship. In Saudi Arabia's harm reduction era, though, there's a much simpler solution: The bins have been covered with heavy lids. The supermarket's trash, now containerized, has lost its luster, rendering the market largely baboon-free.

As I wandered through al-Baha, I saw prominent signs in Arabic warning people to stop feeding the monkeys. The notices, al-Ghamdi told me, have been paired with security cameras and fines of $125 for anyone caught violating the orders. "It's been more effective to target humans," al-Ghamdi said, "but it's expensive." Al-Ghamdi has been involved in efforts to expand walls, fences, and other architectural impediments to prevent baboons from entering homes, shops, and businesses. The trouble, of course, is convincing entire communities of people to change their behavior when even a single holdout—a business owner refusing to cover their garbage, a picnicker hand-feeding clementines to juveniles, or a homeowner failing to put bars on their kitchen windows—can undermine the entire project. Surveillance, even in an ambitious and wealthy autocratic petrostate, also costs money. However, welcoming the world's Muslims to Mecca for the hajj, only to have them pick their way through baboon feces or have their ihram cloth stolen and dirtied on the ground, casts a spiritual stain on the kingdom that cannot be measured financially. For a Saudi regime serving as the steward of this holy place, that means no sum is too great. That's good news for al-Baha, a little over three hundred kilometers south of Mecca, which is already benefiting from this sudden urgency to normalize relationships with hamadryas baboons before they undermine the global worship of Allah. The diminutive mountain city has far less lofty goals than the holy city up the road: a future in which the monkeys are an occasional source of delight rather than a constant threat.

▪ ▪ ▪

Al-Baha and New York City have chosen to target humans to curb their problems with synanthropes, but they're still the exception rather than the rule. In most places, urban animals are still considered enemies to be eradicated. That's understandable when the animal in question is as meddlesome and minuscule as a bedbug (a long-standing synanthrope found in Egyptian tombs dating back 3,500 years). But consider the boar.

In Madrid, the ornate, cramped, and sprawling capital of Spain and the birthplace of Ibérico ham, wild boar are rolling unimpeded through suburbs and satellite towns, causing havoc. There are about forty thousand of them and they have, like so many urban mesopredators, moved from urban neophobia to fearlessness in the blink of an eye. Madrid's boar used to wander timidly in bands of three or four. Now they roll fifteen deep, tearing up gardens, causing traffic accidents, and generally causing mayhem. One wild boar even briefly grounded flights at Madrid's international airport after breaking through a fence and running onto the tarmac. Recently, they've also begun to attack passersby who fail to give them the proper respect.

"Females," says Javier Sintes, "females, they bite. But male ones, they cut with their tusks, and they can provoke very, very serious injuries." Sintes should know. He's been hunting wild boar for decades, and since 2011, he's also been part of a squadron of volunteer hunters who have been tasked by Madrid's network of municipal governments to cull the urban herds. It's a surprising response in a country where guns are highly restricted and a political party called the Animalist Party with the Environment received more than a million votes in a recent Senate race.

"We have seen many, many boars around, and with a rifle and a silencer, we could kill thousands," explains Emilio de la Cruz, a long-time friend of Sintes and as equally ardent a hunter. In one of the most progressive countries on the planet, gunning them all down simply isn't an option, not to mention the issue of friendly fire in urban centers. So Sintes, De la Cruz, and their compadres have found other ways to dispatch the pigs.

"Bowhunting," explains De la Cruz, "is quite efficient." And while it will never net them the thousands of boars they would prefer to kill, it makes up in discretion what it might lose in volume. About 150 nights a year, Sintes, De la Cruz, and about 50 other volunteers are called to suburban parks and fields around Madrid. Once there, they climb up trees, set up stands, and sit overnight in silence waiting for their prey while wearing camouflage and night-vision goggles, readying themselves to rain arrows down on unsuspecting boar, while in the houses and apartment buildings that surround them, *Madrileños* sleep soundly. That is, until the boars, punctured by arrows, start screaming.

"Our goal," says Sintes, "is to scare them. Just to be saying to these animals, 'Here there is no peace. Here, there is a menace.' That's when they begin to go away." The point, he says, is prevention. "Our success is not in killing the animals—it's in the level of fear we put in an area. Boars are clever. When they begin to see that there is blood on the ground, that there are smells of killing around, the herds move to natural areas and away from the urban outskirts." Sintes is also evidently keen to preempt any charges of animal cruelty. "When done in a correct way, with a correct shot, well, it's only around a median of fifteen seconds before the animal loses consciousness," he says. "And then without feeling anything, a median of two minutes of increasing anoxia, without feeling nothing, and they just die there."

Sintes makes it sound peaceful—pleasant, even—but the reality comes into focus as the hunters explain the details. "When you shoot a hunting bow," Sintes says, "if you have hit a lethal point, the animal usually runs for about forty meters." Forty meters—about one hundred thirty feet—may not seem like much ground to cover, but locating a dead boar in the stillness of the gray, streetlight-enhanced Iberian night can take hours, even with night-vision goggles, trails of blood, and dachshunds to help track the dying beast. By the time they find their victim, Sintes and De la Cruz tell me, it's often two or three in the morning, and they still have to haul the boar carcass (which can weigh up to three hundred pounds) onto a truck and deliver it to the municipality's sanitation agency, where it is tested for disease and contamination.

If the animal is healthy enough, it ends up at a soup kitchen. "But that is very rare," adds Sintes. "An urban wild boar that has been, since it was a piglet, eating lots of garbage, and rotting food—those animals, they are normally not edible. Usually, they chop it up and put it in a special spot in a national park where the vultures feed on them." When the meat is too vile even for the vultures, the agency burns the corpses in massive incinerators.

Sintes and De la Cruz are deeply committed to the lifestyle. They work Monday through Friday and would likely spend every night of the week shooting arrows at boars from trees if they could. But Spaniards famously love to party late, especially on weekends, and the combination of a few bottles of Garnacha, synanthropic boars, and trigger-happy bowhunters is too explosive a mix. "We want to be there all the days," Sintes tells me, "as much days as is needed to introduce more fear for the animals, for the menace."

"Right," I venture. "But you also have day jobs, right? You're volunteers?" The two men nod. "So you stay up all night, until five in the morning sometimes, then you go home, you have a shower, and then . . . you go to work?"

"Yeah!" replies Sintes, laughing. He's retired, though he spends his days making nature documentaries and giving lectures on hunting and conservation. De la Cruz is a sales manager at a distribution and logistics company, a position that one assumes would require a good night's rest.

"It's not only the work," adds De la Cruz, "but also our families." They take the sacrifices in stride.

Both men readily admit that killing boar is a short-term fix. After a few weeks, herds will return to the site of the murder and continue as if nothing happened. Despite my efforts, I also couldn't find any evidence that bowhunting had made any meaningful difference in reducing or even stabilizing urban boar numbers across Madrid. But as I reflected on it, it occurred to me that this focus on results might be off target. When I began to think of Sintes and De la Cruz as predatory animals, the scene took on an entirely different complexion. This isn't a failed population-control strategy but an example of a class of predators—bowhunters—drawn into urban areas in pursuit of other synanthropes. It's the equivalent of red-tailed hawks moving into cities to feast on slow-witted pigeons: a sign of ever-greater ecological complexity unfolding in urban habitats. It's also, I imagine, just as thrilling to spot a bowhunter in the urban canopy as it is a red-tailed hawk readying itself for the chase.

Elsewhere, more systematic boar culls are carried out nightly across Madrid through trapping and lethal injection. These must be done by trained veterinarians, who abhor the work and bemoan the need for mass murder when social fixes, like properly disposing of garbage and ceasing to feed the animals, are the only sustainable solutions. Compared to the grim and clinical work of euthanizing boar in the back of animal-control vans, Sintes and De la Cruz's animal-themed expeditions have the feel of playful reenactments of somber scenes, as if they were an extreme form of LARPing. Ultimately, it's hard not to think of urban bowhunting as a symptom of a failure to manage urban ecosystems, though it is undeniably one more way that synanthropes make cities more interesting places to live. In any case, it seems the hunters will be at it for a while. The number of the beast continues to rise, with recent estimates suggesting there could be as many as 2 million wild boars across Spain. Evidently, Sintes and De la Cruz will need a bigger supply of arrows.

▪   ▪   ▪

Containerizing waste is an obvious way to reset a problematic relationship with synanthropes (bowhunting, less so). Sometimes, though, creating urban spaces that promote healthy relationships with synanthropes means more than just better garbage disposal. In Seattle, the potent stew of toxic chemicals leaching into the Pacific Ocean from multiple Superfund sites is a conservation nightmare. Cleaning it up has required hundreds of millions of dollars in environmental repair, but it's proven worthwhile: Despite the marine apocalypse that surrounds Seattle, life—including the giant Pacific octopus—has found a way to proliferate here. But the octopus isn't the only one eking out an existence, nor is pollution the only issue facing marine synanthropes.

"Did you know that baby salmon are afraid of the dark?" It's a delightful non sequitur from Eliza Heery, the marine ecologist who discovered that the giant Pacific octopus is a deep-sea synanthrope. She shares that tidbit as we walk along the beach at Cove 2, mere steps from the water's edge and the anthropogenic debris that lies beneath. I can't help but laugh at the image, but that isn't why Heery wants to talk about it. Across the water from Cove 2 lies Seattle's downtown core and Pier 66, a commercial port and tourist hot spot that welcomes about 2 million visitors per year, who wander and lollygag along its boardwalk, staring out at the brisk beauty of the Pacific Ocean while oblivious to the octopuses, arsenic, and human detritus that it contains. The boardwalk at Pier 66, explains Heery, juts out over the seawall, casting long shadows into the water below. And that has forced the millions of baby salmon born each year here to avoid the shallows, which is where they find food and are protected from predators. Because of the boardwalk's shroud, the fish end up swimming in deeper waters, where they are eaten by harbor seals, black cod, catfish, and smallmouth bass, and where their primary food sources, including plankton, algae, and insects living amid the muck and crevices of the shoreline, are much less abundant. Over time, this has left salmon stocks at risk, which has had a knock-on effect across the urban harbor's ecosystem and on human access to high-quality seafood.

In 2017, as part of Seattle's four-hundred-million-dollar revitalization of Pier 66, marine biologists begged local authorities not only to think about the baby salmon but to think like them too. How could coastal infrastructure be designed to be less scary for the young fish? In a sign of the importance of the urban ocean to Seattle's future, the government decided to indulge the thought experiment. When it was time to rebuild the boardwalk, they agreed to replace the wooden planks and concrete with frosted-glass flooring, which allows warm light to shine down on the water below. The sheer seawall under the boardwalk was also replaced with rounded stones and textured concrete slabs that provide a perfect intertidal ecosystem for algae and insects. Studies done since then have shown that baby salmon have flooded back to the reengineered seawall in greater numbers, where they feed on the abundant nutrients and flourish under the dappled urban sunlight. Listening to the happy babies gossiping under the waves, Melampus would no doubt be proud.

■　■　■

The young field of urban ecology is barely three decades old. As it matures, it's revealed that cities are teeming with animals, along with the subtle ways that urban spaces nudge species into conflict or harmony. Squirrels, case studies in overreaction, are prone to manic behavior on the flimsiest of pretexts. *Why,* I often ask myself after passing a tree and seeing one of the puffed-up rodents staring down at me and screaming, *have you abandoned your chill?* In Baltimore, a couple ecologists tried to answer this question with more rigor. They observed gray squirrels in vastly different neighborhoods across six parks with varying amounts of forest canopy and numbers of trees. They found a clear relationship between the number of squirrels in a park and their level of aggression, suggesting that overpopulation can trigger intra-squirrel conflict. Surprisingly, though, the more trees there were in a park, the less aggressive the squirrels became. What's more, one of the most potent predictors of reduced squirrel aggression didn't have anything to do with the parks themselves. Rather, it was the amount of tree cover in the neighborhoods surrounding the parks. The more tree-lined the roads were, the calmer the squirrels became. It was a lesson both in the importance of city parks as well as in the limits of what any one landscape feature can achieve.

Beautifying our pedestrian paths, public areas, and parks makes our experience of urban life much more pleasant; it's as close to a self-evident truth as one of Soulé's normative postulates. But urban design might matter more to other species than our own. Putting lids on trash cans, building more interesting boardwalks, and planting trees: These are nice-to-haves for humans but critical for our relationship with synanthropes. Still, it's not enough. The Baltimore squirrel study makes it clear that improving a single urban fragment like a park can only get you so far. If it's surrounded by nothing but human-created urban cliff faces like skyscrapers or apartment blocks, concrete freeways, and high-speed vehicles, the squirrels won't stop screaming. The only way to quiet them is to elevate our minds a magnitude more and perceive the city as a single ecosystem rather than as a fractured assemblage of microhabitats.

A city where organisms are relegated to certain places they cannot leave isn't a city of synanthropes; it is a place where animals are subjugated in ecological traps. The wonder of synanthropes, I hope I've convinced you, is in their wildness and freedom. What's beyond debate is that global warming, habitat loss, and intensifying weather events are forcing more species into our cities, where some will make their last stands against extinction. You cannot delay extinction debt by beautifying a parkette. But if you consider the city the way a mesopredator might, as a home range of interconnected habitats to be wandered through, the path forward reveals itself.

Cities are so full of ecological traps that it's unlikely any degree of urban design will unwind all of them. One way to stop a trap from ending a species is to create connections across urban spaces so that animals always have an escape route. Doing so doesn't just hedge against traps but reduces the pressures of both overpopulation and extinction. In an ideal "synurban" environment, each urban fragment (a unit as small as a park or as large as a neighborhood) would respond to the needs of the synanthropes that live there while also serving as a conduit to allow them to traverse the city's entire ecosystem and beyond. In many cities, these green pathways already partially exist as parks, ravines, rail lines, and waterways; the challenge is finding ways to reconnect the segments of the trails that have been severed as cities evolve, while also protecting animals from the dangers they face when they use them. When we create connections across urban spaces, we free animals from ecological traps by easing their passage to other high-yield habitats. This in turn reduces conflict between humans and other animals by allowing them to distribute themselves across less dense parts of an urban ecosystem. Perhaps most important, it also allows for the exchange of alleles across isolated synanthropic populations, extending their relaxation time and staving off extinction.

Some cities have gotten this message. In 2023, Houston, Texas, completed two land bridges over Memorial Drive, a mid-century, six-lane freeway and impassable barrier for animals and human pedestrians trying to cross between two sides of Memorial Park, a fifteen-hundred-acre green space. Prior to the land bridge's opening, park staff dealt with multiple deadly collisions each week, the victims mostly raccoons, coyotes, armadillos, opossums, deer, and snakes. Since its opening, deaths have plummeted. What's more, saving the lives of synanthropes has made Memorial Park a more welcoming place for humans too. The land bridges are stunning architectural accomplishments resembling waves of green gently rolling above a calm sea, while cars pass through the twin tunnels underneath, flooded with geometric sunlight. Off-white pedestrian trails arc across the bridges and throughout the surrounding parkland like tasteful crop circles, leaving the imprimatur of conscious design overlaid upon the meta-intelligence of nature. Under the highway tunnels, culverts were built to move water and animals, a more furtive pathway for those timid synanthropes that prefer to travel through dark and contained areas, which are known as "cover obligates" (as opposed to "openness obligates," animals that prefer wide-open terrain in order to spy their predators and prey from a distance). The overall impression is of the serenity and connectedness of nature alongside the comforting calm of human design.

Elsewhere, land bridges and other wildlife crossings are becoming more widely adopted to preserve and grow urban wildlife populations. The state of Colorado is among the world's leaders, having built twenty-eight major game-crossing structures spanning fifty miles of mountainous freeways along with four hundred miles of fencing to guide animals toward the safe routes. It's an audacious undertaking but the results are undeniable. After the wildlife crossings were constructed, Colorado experienced a 90 percent reduction in collisions and wildlife carcasses along its state highways. In other places, like Canada's Banff National Park, where plains bison and grizzly bears roam, the creation of a network of forty land bridges dropped collisions by 80 percent. The Memorial Park land bridge in Houston and the forty-five-acre restoration that accompanies it cost taxpayers $70 million. The total cost of Colorado's network of wildlife crossings, culverts, fencing, deer guards, and highway improvements, all designed to re-connect habitat fragments, isn't publicly available, but each crossing costs somewhere in excess of $10 million, making this an investment in the hundreds of millions of dollars. Banff's system has cost roughly $90 million since its inception in the 1990s. It's an inspiring commitment and evidently results in fewer deaths. But the sheer cost of these ventures raises a question: How do you motivate that level of change among skeptical urban dwellers when so many other human priorities scream for attention?

One approach is to rely on charismatic megafauna to sell the project for you. In Los Angeles, a city built on celebrity, that honor fell on P-22, the mountain lion whose movements became an episodic soap opera followed by Angelenos and audiences around the world during the COVID-19 pandemic, transforming him into the season's hottest synanthrope before his death in December 2022 after a collision. The collective mourning was immediate and intense. A mural was painted in his honor at the Griffith Park Visitor Center. News outlets the world over covered his death as they would a member of the British royal family ("Requiem for a Great Cat," read the obit headline in *The New Yorker*). And his brief but adventure-filled life was adapted into a series of books, including a kid's book called *The Cat That Changed America* and a novel, *Open Throat,* narrated by P-22 himself.

Books and a mural are nice legacies, but P-22's life and death did more. Genetic testing revealed that he had been born in the Santa Monica Mountains, part of the dwindling population of mountain lions hemmed in by U.S. 101. Defying all laws of probability, P-22 crossed both U.S. 101 and Interstate 405, a twelve-lane freeway among the busiest in the country, managing to exit the dwindling home range and flee the omens of his brethren's looming extinction debt: kinked tails, testicular anomalies, and low fertility rates. The genetic withering away of the mountain lions was like a slow-motion car crash, and efforts had been ongoing since 2015 to build a wildlife crossing over

U.S. 101 to allow for them to exchange alleles with another group of mountain lions that lived north of the freeway. Until P-22 became a celebrity, though, every effort to fund the project stalled out; people simply didn't care enough. At $90 million, the project was a pipe dream in a city saddled with rampant wildfires, widespread homelessness, and seemingly endless infrastructure challenges. But as his light shone brightly and then faded away during COVID-19, the realization that Los Angeles was home to wild and beautiful creatures suddenly became a source of pride, and P-22 became the manifestation of a collective id, a creature free to indulge its freedom while the rest of us cowered at home. By April 2022, after years of disinterest, P-22's travels had inspired donations of roughly $87 million from more than five thousand individuals, foundations, businesses, and nonprofit agencies. In the fall of 2026, the Wallis Annenberg Wildlife Crossing will finally be unveiled: an acre-wide vegetation-covered bridge surrounded by twelve acres of restored habitat, which will make it the single largest animal crossing in the world and the best last hope to save P-22's kin from extinction. It's the perfect Hollywood ending: the grizzled hero sacrificing himself so that others can live.

▪ ▪ ▪

It starts and ends with connectivity. Being connected to animals makes our lives richer. Building connections for animals between fragmented urban ecosystems makes cities safer for all the species that inhabit them, and more beautiful too. Corridors like trails, culverts, and land bridges have the highest upside in supporting urban species, but their cost and scale mean they're often infeasible. Animal paths map poorly over city grids, and there's little appetite to radically reconstruct them in nature's image.

In response, urban designers have increasingly turned to "stepping stones," which are disconnected green spaces set across cities like dry rocks across a creek. These habitat islands are refuges for synanthropes, and they're most effective when placed between larger habitat patches like forested parks and meadowland. The beauty of stepping stones is that they can be built into pretty much any existing infrastructure at any scale. In New York City, the Greenstreets Program has transformed paved and vacant traffic medians into tiny stepping stones for birds, insects, and small mammals (yes, including rats). It gives synanthropes a chance to find shelter and food as they crisscross the city on their way to larger and friendlier habitats like parks. There are human bonuses too: The vegetation in these microhabitat patches reduces the urban heat island effect while the ground cover absorbs stormwater, reducing flash floods. They look pretty too. Where there used to be concrete slabs, there are now riots of flowers, trees, and greenery, surrounded by pollinators and birds, rising amid the clamor of Manhattan streets.

Other cities have taken the idea of stepping stones to preposterous heights. In Singapore, a tropical city-state of 6 million people crammed into an area about half the size of London, space is at a premium, so the idea of carving out residential areas to build animal corridors is laughable. But as a small island country that has expanded at the expense of a tropical forest (only 0.5 percent of which remains), creating sustainable urban infrastructure isn't optional. The city-state experiences extreme urban heat island effects, with an average temperature 9 degrees Fahrenheit higher than surrounding areas. Being an island, there is also nowhere for people to go if pollution gets out of control or sources of fresh water run out.

Singapore's answer has been to build the city as high as possible to accommodate a growing population while transforming the skyline into a series of tightly packed stepping stones. Skyscrapers are covered in palm fronds, long grasses, orchids, ferns, and hundreds of other plant species that run across roofs and balconies and climb vertically down to the ground, creating a sense of continuity between

AI-powered urban computing systems, freshwater capture, and nature's joyful anarchy. Two-thirds of Singapore (including its sidewalks) has become one big water catchment system, funneling fresh water into underground reservoirs. Leaning into nature as a partner has also inspired the city toward delightfully bizarre architectural accomplishments. This includes massive conical buildings resembling dormant volcanoes swirling to the ground, entirely covered in a motley pastiche of plants. In the Gardens by the Bay, a seaside nature reserve, a dozen "supertrees," each 150 feet tall, cast sprawling shadows across the acreage. These concrete-and-metal fabrications are designed to generate solar power, vent heat from nearby conservatories, and redistribute rainwater into the city's reservoirs. Their long branches create a canopy, bringing urban heat down, and they are teeming with life: Close to 200,000 different native and exotic plant species have been set into their trunks and stems. It's an odd juxtaposition: an autocratic and hyper-ordered city-state that has immersed itself in the unending possibilities of life, remixing and synthesizing the natural world into striking new harmonies. Meanwhile, synanthropes have come along for the ride. The city is replete with house crows, rock pigeons, Javan mynahs, and macaques, all of which find solace in stepping stones that span all three dimensions of the urban matrix. Tilapia have returned to the city-state's rivers now that the water is free of pollution; this has in turn attracted smooth-coated otters, a singular synanthrope that has ably filled an ecological vacuum that coyotes, boar, and raccoons fill elsewhere. Finding a niche in Singapore has turned the docile and timid otter into a classic urban mesopredator—devoid of neophobia, highly territorial, and even prone to biting humans.

Singapore's otters are evidence that nature's chaos can never fully be contained, even in places that seem like synanthropic utopias. Still, the problem of aggressive mustelids pales in comparison with the monumental challenge of bringing order to human dystopias.

▪ ▪ ▪

When Dharani was killed by a Bengal tiger, Karuna Mandal's life was destroyed. There was no point in registering his death with the state: Like almost all the fishermen in the Sundarbans, Dharani had entered the tiger reserve illegally. Instead of the 500,000 rupees in compensation the government advertised as recompense to families whose members were killed by tigers, she was given a stipend of 1,000 rupees a month—less than forty cents a day.

Mandal's story is a personal tragedy. It is also one of more than three thousand stories told by women in the Indian Sundarbans who have lost husbands to tiger attacks, though interviews with village elders across the densely packed archipelago suggest that the true number could be six thousand or more. The ranks of tiger widows appear to be rising in step with the greater desperation of the people here to scratch out a living as climate change causes their islands to recede, a problem that has seemed simply too hard to solve.

Or it did, until a local high school teacher faced down the loss of his entire community firsthand. In May 2009, the Sundarbans was pummeled by Cyclone Aila, a seventy-mile-per-hour weather system that ran directly north over the Sundarbans before tearing into the heart of Kolkata. Over three terrifying days, the cyclone left 1 million people homeless and almost 350 dead, while causing hundreds of millions of dollars in damage.

Umashankar Mandal had lived his whole life on Satjelia Island, at the edge of the tiger reserve, happily teaching geography to eager students. After the storm passed, he emerged from his home to see Chargheri, his community, torn apart. Mud homes had been flattened. The small plots of land that farmers used to coax a meager bounty from the soil and feed their herds had been covered in salt from the brackish waters. Thousands required medical care, but none was available. The island itself looked as if a great claw had raked across it, destroying young mangrove stands and severing hundreds of feet of coastline, which were thrown under the silty waters of the Garal River. (In the immediate aftermath of the cyclone's damage, Sonagachi, Kolkata's main red-light district, saw a 25 percent surge in the number of sex workers arriving there, many from the Sundarbans.)

The devastation radicalized Mandal. As a geographer, he understood rationally how vulnerable the region was to climate change, but it was entirely different to experience it himself. To clear his head, he went down to the shore and looked out at the opaque and turbulent Garal River, where hundreds of green buoyant discs were bobbing in and out of the rushing water. These were the seeds of *Avicennia germinans,* the black mangrove, which germinate before falling from their parent tree, allowing them to spread across vast distances and begin growing immediately once lodged in the clay-like ground. This evolutionary adaptation, called viviparity, is how the Sundarbans became the world's largest mangrove forest, a green shield capable of dampening the worst effects of cyclonic winds for tens of millions of humans. In an age of illegal logging, climate catastrophes, and rising sea level, Mandal also knew that the mangrove forest was slowly being eaten away, and that nature's resilience wasn't going to be enough to protect him and his community from the next apocalypse. So he headed home, grabbed a net, and returned to the mudflats. Then he waded into the river, catching dozens of the germinating mangrove seeds, walked carefully back through the mud, feeling his legs sink to the knees, and began planting them in the ground. And then he did it again. Within two months, Mandal had planted thousands of mangrove seeds along the barren mudflats of Chargheri and other nearby communities on Satjelia Island.

That was more than fifteen years ago. Since then, Mandal and his supporters have planted 1 million black mangrove seeds across this region. He's also become increasingly audacious in his goals, each summer recruiting hundreds of volunteers to help him collect mangrove seedlings floating along the many rivers of the Sundarbans, and to plant them in small reusable baskets before securing them in the mud. No expert in One Health, Mandal has nevertheless become one of its greatest practitioners, acting on his belief that the health and resilience of the mangrove forest is critical to protecting the people and animals that live here.

Mandal, a man of few words, wears a beatific smile and moves with a gentle and unthreatening elegance. He is instantly likable, a person who has committed himself to a cause that is self-evidently good. If Soulé were alive, I am sure he would see his own beliefs at work in Mandal's decades of environmental reclamation; his obsessive labor is Soulé's crisis discipline in action. For all that, Mandal is also something of a rebel, bucking the deeply held beliefs underpinning this region's social system, which has long considered tiger widows inauspicious beings to be cast out. Mandal sees something else entirely in this growing legion of women: an army.

Over the past few years, Mandal has used the success of his mangrove planting project to reclaim not only land but also the social connections that fall away once a woman's husband is killed by a tiger, leaving her destitute and isolated. Under the auspices of the Purbasha Eco Helpline Society, a charity he founded to support his ecological work, Mandal has brought tiger widows together as a hundreds-strong "mangrove army" tasked with expanding the green shield of the forest to protect communities. "The biggest impact is in restoring ecological balance," he tells me, "but it's also created more fishing grounds, bee cultures, and provided carbon sequestration." Because of the high frequency of extreme weather events here, he's seen the benefits of his efforts many times over. "When the cyclones hit the landmass," he says, "our green shield is there. Before, all we had was an embankment made of concrete, which doesn't protect anyone. But the mangrove plantations restrict the severity of the wind, and they stop the bank from eroding due to water surges following cyclones."

There are smiles all around as tiger widows gather barefoot on the shores of the Garal River wearing neon-green construction vests with the words MANGROVE ARMY written in all caps across their backs. They spread themselves across the mudflats in the humid summer heat, some wearing their saris over their heads for protection. Their movements are slow and deliberate, some carrying tin baskets of foot-long mangrove saplings from a *chandi* maneuvered into the shallows. From there, the baskets are slid along the sticky mud and distributed across the members of the mangrove army for planting. Walking on mudflats is a practiced art here, and it is immediately evident when someone hasn't contended before with the sticky clay-like substance. (The trick, Mandal tells me, is to walk barefoot and put more stress on your toes than your ankles, though even people who grew up here still get stuck all the time.) As the heat rises throughout the day, a constant volley of the Sundarbani dialect of Bengali settles over the action as women, some kneeling or half sunk in mud, secure the twig-like saplings. It is tough work but done together, eliciting a sense of collective ownership among the women who spend their days expanding the mangrove forest and ensuring that the island they live on will remain above water, at least for a few more years.

Mandal has become known in the region and internationally as the "Mangrove Man of the Sundarbans," with his Purbasha Eco Helpline charity receiving donations from the World Wildlife Fund and others to expand the plantations. Beyond bringing tiger widows together to build the green shield, Purbasha Eco Helpline Society offers them basic food supplies to help them weather their circumstances. Karuna Mandal, for instance, is given a regular supply of groceries, mango and water apple saplings, and country chickens, the latter of which can provide a source of meat, eggs, or money.

Since 2009, Mandal's vision has grown exponentially. Replacing lost mangrove habitat is still the central task of his life, but he is now working to rewrite the social ecology of Sundarbans. It's a curious reversal of synanthropic evolution, with humans forced to adapt to life at the precipice between urban and wild spaces. Like a species in the throes of extinction debt, tiger widows are too socially isolated and economically vulnerable to survive without reconnection. They are also often forced to continue the risky business of harvesting fish, honey, and crab from the forest, which leaves them destined to become tragic understudies to their dead husbands. It doesn't matter how devoted each widow is to Bonbibi, or how closely she follows the tenets of only taking what she needs from the forest. As long as there are 150,000 people chipping away at the fringes of Dokkin Rai's kingdom, neither the tiger lord nor those under the protection of his nemesis will survive. The key, then, is to wean people away from their reliance on the forest. Tiger widows are among those most willing to do so, having little left to lose, and Mandal has learned that it doesn't take much to nudge them down a different path. Many of the women I spoke with were given seeds (okra, pumpkin, spinach, and legumes) every few months by the Purbasha charity, along with agricultural training to get the most out of the uncooperative saline soil. Mandal also organizes vocational training, especially tailoring and dressmaking, and has distributed sewing machines to tiger widows so they can build their financial independence. He's also dredged hundreds of ponds across the island communities to serve as small fish farms, making widows and other forest workers less reliant on barely there government support or the dangerous riches of the tiger reserve.

Mandal is keenly aware of the spiritual resonance of his efforts, particularly the act of reclaiming destroyed coastline for the mangrove forest. "We are trying to build a relationship between women and the plants," he tells me. "In 2025, Purbasha started a program where women tie a *rakhi* to the trunk of mangrove plants." It's part of an annual Hindu rite known as *raksha bandhan,* held in August each year across India to celebrate bonds of love. *Rakhi* knots symbolically bind two people together, protecting them both while ensuring that the relationship lasts for a lifetime. "So here," says Mandal, "women tie the knots onto the mangrove saplings and their own wrists, to protect the plants from any further damage caused by natural calamities and to protect themselves while they are working on these efforts." Among women who have lost so many of the ties that bind them—their husbands, their families, their communities, and their livelihoods—reclaiming a connection to the land is a powerful act.

It also plays an important ecological function. The mangroves are so well adapted to bringing order to this fluid, brackish, and chaotic ecosystem that it takes only one year before they can protect the villages from high winds and erosion, their intricate root systems holding embankments in place, their twisting branches and velvety leaves creating drag that cuts down the force of winds. This protection is invaluable when a cyclone hits, translating into countless lives saved and communities spared total destruction, not just across the 4.5 million people who live in the Sundarbans but among the 15 million just north of here in Kolkata proper. Without 10,000 square kilometers of mangrove forest a few dozen kilometers south of the city, the damage would be even more apocalyptic: In some projections, many of Kolkata's most densely packed neighborhoods could be submerged if the islands of the Sundarbans continue to erode. As it stands, the Sundarbans has lost 440 square kilometers to erosion since 1991. Families have been displaced as successive mud embankments get subsumed. People have had to rebuild five times or more as the coast shrinks, the location of their original homes sometimes as much as a kilometer out from the shore, a storehouse of memories drowned beneath the waves. But each mangrove sapling planted by the mangrove army fixes the islands in place a little more, in turn making it less likely that a widow will lose her home in a storm and be forced to flee to the darkest parts of Kolkata's economy. It is nature as the last and best protector of humanity, Bonbibi made manifest in the tangled roots holding the muddy land together in a loving embrace, the knots of the *raksha bandhan* tighter and tighter around the mangrove trunks each year as they grow into the future forest.

Call it One Health, where animal, environmental, and human health collide. Call it the rules set out by Bonbibi and Dokkin Rai, in which the forest must be protected and respected at all costs. Or, in the language of global philanthropy, call it a micro-grant and upskilling project to lift vulnerable women out of poverty and end the cycle of sex trafficking and child exploitation. These are different names for the same story: navigating a peaceful path through the most brutal edges of synanthropic conflict.

■ ■ ■

I wrote that the protection from the mangrove forest is invaluable, and it is. That doesn't stop ecologists from trying to calculate its exact monetary benefit, as well as the value of every other component of the natural world. The mangrove forest in the Sundarbans, they say, is worth $543.3 million in savings when a cyclone hits the region, and the region lost more than $3 billion in value over the last forty-five years on account of erosion and habitat destruction. This way of speaking about nature is known as "ecosystem services," and it's become the dominant narrative in ecology. It's also a direct rebuke of Soulé's normative postulates. Applied to the real world, Soulé's description of the self-evident goodness of diverse and thriving ecosystems ends up turning most human beings into villains simply because they rely on the natural world to live (incidentally, it also turns a small number of mostly Western biologists into nature's saviors). In extreme cases, like those of the tiger widows, Soulé's normative postulates break down completely. Karuna Mandal's late husband, Dharani, isn't a villain because he illegally entered a forest sanctuary, nor is she for having survived off of his catch. They are as much the victims of ecological imbalance as the tiger that killed him. According to the normative postulates, though, Karuna's lack of commitment to the idea that the needs of nonhuman species outweigh her own makes her part of the problem—one that must be dealt with as efficiently as possible, in line with the urgency of the crisis discipline.

When Soulé's purity of intent collides with cities and the synanthropes that inhabit them, it can lead to unworkable solutions. Though cities are on balance more biodiverse than the areas that surround them, you cannot simply deny the needs of the human beings who live there, as seductive an eco-fantasy as that might be (wouldn't it be lovely to replace all roads with footpaths that we share with coyotes, raccoons, foxes, baboons, and snakes, as ravens and hawks caw at us from above?). That type of thinking leads to "fortress conservation," the forced eviction of human beings from ecosystems, and in some cases even the denial of Indigenous rights to land stewardship in pursuit of absolute habitat protection. The Sundarbans is a textbook example of how fortress conservation can go sideways and what happens when it does: higher levels of illegal resource extraction after the imposition of forest guards, increased vulnerability to ecological threats among urban local communities, greater mistrust of government, and more death—quickly and soon by a ravenous tiger, or later and slowly as islands erode and communities are erased. It's also led to increasingly vocal calls for a "managed retreat" of all 4.5 million inhabitants of the Sundarbans to protect the area from human destruction. This is hardly a harmonious future.

The antithesis to Soulé's postulates is the language of ecosystem services. In this framing, humans are placed squarely on top of the natural world, with nature as a menu of amenities with fixed price tags that our species can choose to benefit from. Eliza Heery has been steeped in this language since she entered the field and describes it to me as "the monetization of conservation science," in which every single aspect of the natural world can be quantified. It's so ubiquitous among ecologists that most simply take it for granted.

As urban areas have become central to the survival of countless species, ecosystem services has become the go-to way that ecologists articulate why we should care about creating healthy habitats. It's a savvy strategy to sell urban renewal to policymakers who must back up their decisions by appealing to costs and revenues. Soulé's refrain that "biodiversity is good" isn't going to get you very far in a city council meeting, but mentioning that the global value of all ecosystem services—a.k.a., nature in its entirety—is $125 trillion a year might motivate some councilors to muse about how they too could get in on some of that cash.

To me, the ecosystem services framework is the height of cynicism. It's predicated on the notion that the only way humans can fathom the self-evident goodness of nature is by asking, "What's in it for me?" It's nature as commodity, its value as relative as the latest meme coin. When I speak with Heery, her language is peppered with references to the ecosystem services that octopuses provide, even as she critiques this framing of how humans relate to nature. So far, though, no other approach has taken its place. "Soulé's normative postulates were conservation science 1.0," she tells me. "Ecosystem services is 2.0. We're now entering 3.0. But it's only just forming. It's up to the next generation to create what it's going to mean."

Whatever comes next, synanthropes should lead the way. Set aside arguments about the moral imperative to protect the planet or the services that nature provides to humans, however compelling those might be. There's a much simpler starting point for discussions about how we relate to the natural world: All we have to do is look, listen, and feel for what's living in our homes and neighborhoods. We've created urban centers to dislocate us from the natural world. The best part about them, to my mind, is how badly they've failed to do that. In cities, towns, and suburbs, we are positively surrounded by variegated forms of life, and not just by species we've deemed acceptable. There are snakes and bumblebees, langur monkeys and spotted hyenas, goldfinches and house mice, red foxes and red-tailed hawks, capybaras and coyotes, and an endless assortment of synanthropes that, despite their differences, all share the burden and opportunity of living with *Homo sapiens*. Let's build a language recognizing that cities are truly unnatural places and then celebrate the menagerie that nevertheless survives in them, dispensing with the nostalgia we have for a lost and pristine world.

If we can peel ourselves away from the stultifying grief of climate collapse, there is real joy to be found. I'd also argue that building a conservation movement based on joy will pay greater dividends than one that moves to the tune of a funeral dirge. Aside from insects, birds are the wild animals city dwellers are most likely to encounter. They're also far more abundant in urban areas than in wild ones. And while there are many studies outlining the ecosystem services that birds provide, none can explain why we love to hear them sing. Birdsong is the sound most people report best relieves their stress, and these benefits multiply exponentially as humans hear the songs of different species overlapping in a chorus. The reasons why, I think, are connected to deeply ingrained spiritual, emotional, and biological responses that are far more complex than a mechanistic understanding of their economic value. Perhaps the sound of birds happily chirping is a long-standing signal that we've arrived in a rich and fecund habitat, making it the sonic equivalent of petrichor—the smell of rain falling on dry soil—which humans are uncommonly able to detect at levels as low as 0.4 parts per billion, a trait that would have saved our prehistoric ancestors on their long marches across the globe. I'm loathe to overthink it, though; it's enough that we love to hear birds sing. We love the clicks and cries of ravens, our Paleolithic companions, and the efficient chirps of house sparrows too.

But our love does not end with animals that pose no threat. I was in awe of the sound of a lusty coyote pack howling across the canyons surrounding the Oaks in Los Angeles; we are stunned into silence by leopards roaming through courthouses and residential neighborhoods. The reason we love even the most threatening specters is that human aversion and fascination are intertwined. Over time, we distill our fear of the power of predators into stories and dreams, and these become the material informing the rudder culture guiding all human societies. Edward O. Wilson, the famed American naturalist, called it biophilia: the innate human affinity for living creatures, which is always tinged with an underlying terror of what it reveals about us.

As we transform our cities, we will no doubt have to quantify our successes, and that's not a bad thing. "Ecosystem services has a role to play," Heery tells me, "but only if we are hyperconscious of its limitations and bias." Accounting for the value of nature isn't inherently problematic. The issues arise when we conclude that what we're quantifying is all there is to know. To keep us out of that trap, I'd offer an alternative: Let's approach ecology as we do cosmology, the study of the universe. When cosmologists glimpse a dying star, they do so indirectly through the light and molecules that it's emitting. The conclusion isn't that we've seen all there is to see; it's that we've added an infinitesimally small piece to our knowledge of the vast unknown of the universe. Each cosmological discovery—from gravity to the speed of light to black holes to quantum particles—reminds us how little we know. When we look to the stars, we are confronted by beautiful questions. So too, I say, for synanthropes and cities. Quantifying that bats provide $22 billion annually in pest control in U.S. cotton fields gives us a defensible rationale to deploy resources to stop white-nose syndrome, a fungus that has attacked leaf-nosed bats and caused their numbers to plummet. But will that number make us love bats? No. For that we must go to where they roost and witness them launching themselves into the sky at dusk, twirling bundles of fur on impossibly thin wings, speaking in languages beyond the acoustic limits of human comprehension. Let them draw ever closer to us in fearful intimacy, one variation among many revealing the mystery and desperation of life in all its forms.

For those who are lost, there will always be cities that feel like home.

—Simon Van Booy, *Everything Beautiful Began After*

In 2008, sixty years after its inception, the International Union for Conservation of Nature made a curious announcement. An international agency tasked with identifying species at risk of extinction, the Union has long been the tip of the spear in calling attention to the imminent loss of wild animals and plants through its Red List of Threatened Species. The headline in 2008 was grave: One in four mammal species was at threat of extinction.

Beyond the headline, the Union quietly weighed in on one mammal that had long been neglected in its taxonomy. After sixty years of scouring the planet's flora and fauna to systematically assess the risks they faced as Earth went through a ruthless climactic reordering, the group finally turned their attention to the creature at the center of it all. *Homo sapiens,* they declared, had become "a species of least concern."

It was an odd bit of bookkeeping. A species of least concern is the nomenclature for a species that is proliferating well enough in the climate nightmare that no special attention is needed to keep it from going extinct. The Union also placed our species among a motley crew that included houseflies, pigeons, cats, dogs, mice, grasshoppers, common weeds, and all manner of bottom feeders and vermin throughout the world. "There are currently no major threats to humans," the Union noted in its entry. "At present, no conservation measures are required."

It was laughable, a joke even. But peel away the layers of taxonomic bureaucracy, and at the root of the designation lies an important truth. Despite how special we may feel, humanity is just as common as a housefly. Still, there's a path to becoming the rarefied creature we believe ourselves to be: It's in the company we keep.

I'm someone who has always been fascinated by animals, but I'll admit that I've had less contact with them than I'd like. My Instagram feed is full of them, my algorithm very well trained to serve me up the latest in bat, insect, monkey, and manta ray content, and it's hard to pull myself away. I can stare at that screen forever, feeding my thirst, but true intimacy requires entering the real world. One knock-on effect of writing this book has been that it's moved me away from my phone and full-bore into the lives of the animals that surround us. I've shared some of the more dramatic moments along the way, but some that floored me the most were small and by and large don't make for great prose. I've listened to birdsong. I've followed bees through gardens. I've watched coyotes watching dogs and thought about how they must see their domesticated cousins—as silly but pretty creatures oblivious to the prison of their minds. I stared out the large window in the hospital room where my daughter was born, dumbfounded that the drab downtown urban canopy was completely devoid of stepping stones, and lamented its lack of green roofs (I kept my thoughts to myself; there were other things going on). On nights she did not sleep, I wandered around my neighborhood and kept coming across the same skunk pawing at the same patch of soil under a bird feeder from which seeds had fallen. These moments nudged me into greater intimacy with the natural world, and they all happened in the heart of cities.

You do not need to lock eyes with a baboon to feel wonder; all you need to do is follow animals as they move. Once you do, you'll unlock a world of desire paths—informal trails created by the repeated footfalls of animals—that reveal an alternate pattern of movement in cities. Once you look, you'll see desire paths at the edges of parks and through the middle of well-planned city grids. All animals use them, humans included. They are hints of wildness in ourselves and our familiars, revealing that other ways of living are possible. And they are reminders that we will never rid ourselves of nature's chaos, nor of our own.

I've also found a surprising solace in glimpsing the lives of synanthropes over the last few years. In Seattle, when I dove down sixty feet and met a giant Pacific octopus clutching hundreds of thousands of eggs in a house made of concrete and rusted iron debris, a few short months before her death, I felt all the big emotions that you would expect. Running my eyes over her fleshy pink tentacles, white suckers, heaving gill, and gigantic eyeball, other thoughts came to mind that I'm only ready to admit to now. *This thing is a monster,* I thought, *but somewhere in the world is another one like it that feels, with every fiber of its being, that this is the sexiest creature to ever exist.* That thought made me laugh, which is hard to do underwater. And it's as true of an octopus, I realized, as it is of a corpulent raccoon, a hideous-to-me leaf-nosed bat, and a hamadryas baboon flashing a swollen, garish, and throbbing red bum. And if you're waiting for it, there's no lesson here. It just felt revelatory that every creature, no matter how bizarre or diminutive, can produce extravagant emotions in others. I like that. It feels like hope.

Derek Jarman, the filmmaker and writer, was a singular voice who charted a path to joy in mortality as he navigated the AIDS epidemic ravaging London's gay community in the 1980s. It was the era of Margaret Thatcher and Ronald Reagan, of austerity and poverty, when nuclear war, absurdist government propaganda, acid rain, and habitat destruction seemed ready to obliterate the future. But even as he lay in hospital, sick with AIDS, the odds of the survival of the human race getting close to a coin flip, Jarman refused to give in to despair. Instead, he built a garden at the edge of a nuclear power plant. It was a protest, a requiem, and a love story for the natural world all rolled into one. Planting flowers and nurturing growth amid climate and political chaos was for Jarman a radical act of joy. He called the project Modern Nature, and in his book of the same name, he wrote, "[m]y garden's boundaries are the horizon." I can think of no better mantra for the Synanthropocene. Let us stretch the city's boundaries to the horizon and build a society in which all of us—*Homo sapiens* among many—find balance and joy in the other.

Author's Note

When you make animals the main characters in your book, you're bound to run up against some challenges. When those animals are scattered across the world, doubly so. I wrote my last book, *The Invisible Siege,* during the pandemic. It was an odd experience: The book is about how the coronavirus family evolved over eons and about how scientists around the world had been preparing for a situation like COVID-19 for decades. Like everyone, I was stuck at home, so every bit of color for the book came from video calls, peer-reviewed studies on virology, and images of faraway places I could find online. Ultimately, it worked out well enough, but I promised myself that the next project would force me outside to write a book literally from the ground up.

Cue the many animals, people, and places I was lucky enough to commune with on my way to creating the world of synanthropes that you and I have just shared together. Doing so involved working with three basic materials. The first was a deep dive into the scientific literature about synanthropes, urban design, and animal biology, which involved synthesizing about six hundred peer-reviewed studies and then translating them into a language that you and me could understand intuitively. The next was spending long hours with humans whose lives were enmeshed with urban animals, many of whom you met in this book, and finding how to shape narratives around their lives that illuminated the information I wanted to convey about synanthropes. The third was, of course, my time with animals in many different cities, which made up experiences that became the necessary heartbeat of this book.

As I alluded to earlier, it wasn't always straightforward. In some cases, the best-laid plans ended up collapsing unexpectedly, making me rely on other tools to create the world of the book. When I planned my research trip to visit with tiger widows in the Sundarbans, for instance, I was only dimly aware of the chill in diplomatic relations between Canada, my home country, and India, owing to the assassination of a Sikh separatist leader on Canadian soil by Indian government operatives. When I applied online for a visa to let me into the country, which I was assured was essentially a formality, it was immediately denied without official explanation, though it quickly became clear that I wasn't the only Canadian citizen facing these kinds of hurdles. This led to a byzantine and Kafkaesque bureaucratic nightmare that never got resolved. To overcome this complication, I leaned on the techniques I honed writing nonfiction during the pandemic. I was helped enormously by a support team in the Sundarbans led by Debaroti Das, who diligently set up remote interviews and videoconferences with tiger widows and other interesting people, who spent days taking pictures and videos of the region and its people, and who variously provided me with all the material I needed to bring the section to life without physically being there. Sometimes, even when I was able to get to the places I needed to go, circumstances conspired to block my path. That was the case at the al-Baha dump, where upon arriving with Ghanem al-Ghamdi, we were confronted with locked gates and a staff that refused to let us in, despite al-Ghamdi's many previous forays into the bizarre new world evolving amid the trash heaps. What was initially a disappointment, though, became a lucky break. Unable to enter, al-Ghamdi pondered an alternative, eventually remembering that he had taken some videos that might be helpful. He searched his office later that afternoon and the next day met me with extensive footage taken by drone of the dump's canid and primate inhabitants. It was remarkable. The footage accomplished two things that became crucial to the chapter: First, it gave me a bird's-eye view of the scale of the animals' numbers that I would not have otherwise fathomed simply by walking around the area. Second, when the video was taken, the drone's buzzing panicked the baboon horde, causing the stampede that I describe in the book, revealing the role that feral dogs had come to play in hamadryas society.

I share these unexpected moments with you to shed some light on process. It's also to emphasize that nonfiction writing, for me at least, requires immersing yourself so deeply in a world that you are forced to give up control. Those are the moments that make the best writing, I think; they're also the moments in which I feel most alive.

Acknowledgments

This book came to life because of the vision shared between Kirby Kim, my agent, Kevin Doughten, my editor at Crown, and myself. Their passion for and belief in the world of this book were the catalysts that led to its creation. Others were instrumental in shaping this project as well: Lynn Henry, my editor at Knopf Canada, provided excellent guidance, as did Jess Scott at Crown and Amanda Betts at Knopf Canada. Eloy Bleifuss and Lansing Clark at Janklow & Nesbit were diligent and generous sources of support. Natalie Blachere (production editor), Heather Williamson (production manager), and Dianna Stirpe (copy editor) at Crown made this book beautiful and excised as many of my errors from it as possible. No one would have heard of this book without the dedication of Hannah Perrin, who marketed the book (Crown), and the work that Lindsay Cook (Crown), Shona Cook (Knopf Canada), Emi Battaglia (Emi Battaglia PR), and Hailey Dezort (Paratext) did to publicize it. As a former book publicist, I can relate to its triumphs and travails; the fact that they care means so much to me. Finally, I was blessed with wise and ruthless early readers in Josh Cockerill and Aubrey Nealon, whose friendship did not stop them from giving it to me straight.

I talk about this book as a world not only for what it reveals about the hidden ecology of synanthropes but also because of the wide community of people who contributed to it. This includes many of those who helped me, in all kinds of ways, to connect with and understand animals, and without whom this book would not exist: Maddy Curry, Katalin Szlavecz, Dawn Scott, Gal Nissim, Andrew Budziak, Antonia Hammer, Sarah Gunarwan, Debaroti Das, Abdulaziz Alzahrani, Ben Perreau, Lauren Moretto, Gerry Hans, Dietmar Zinner, Jon O'Barr, Earl Norton, Mike Perez, Matt Hannam, Fanny Singer, Jeff Lee Petry, Adam Azimov, Ellen Wong, and Heather Goodchild. I also benefited greatly from reporting done by others, including Craig Silverman at *BuzzFeed*, Julie Davidow at *University of Washington Magazine*, Christina Larson at the *Chicago Sun-Times*, Gretchen Kell at *UC Berkeley News*, and Adam Mahoney at *In These Times*. I also acknowledge support from the Banff Centre for Arts and Creativity for protected time at the Leighton Studio Residencies as well as the Canada Council for the Arts for support in the early stages of developing this book. Three cats played roles in this book: Kenneth, Jason, and my own, Beannie. My parents, Ron and Elly Werb, instilled in me a deep love of art and writing, sometimes to their chagrin, for which I am very thankful. My siblings Justin and Jessica Werb taught me to not to take myself seriously, which is a gift. And finally, I would not have written this book or done much of anything at all without the love I share with Miranda, Zuleika, and Anni.

Notes

Prologue: Death and Desire

4   **its cunning and mystic powers:** S. T. Hussain and C. Baumann, "The Human Side of Biodiversity: Coevolution of the Human Niche, Palaeo-Synanthropy and Ecosystem Complexity in the Deep Human Past," *Philosophical Transactions of the Royal Society B* 379, no. 1902 (2024): 20230021.

4   **mammoth, reindeer, and other big game:** C. Baumann et al., "Evidence for Hunter-Gatherer Impacts on Raven Diet and Ecology in the Gravettian of Southern Moravia," *Nature Ecology & Evolution* 7, no. 8 (2023): 1302–14.

6   **a pathway to animal cohabitation:** Hussain and Baumann, "The Human Side of Biodiversity."

6   **were assembled into pendants:** P. Wojtal et al., "The Scene of Spectacular Feasts: Animal Remains from Pavlov I South-East, the Czech Republic," *Quaternary International* 252 (2012): 122–41.

6   **at least one species of crow:** J. M. Marzluff and T. Angell, *In the Company of Crows and Ravens* (Yale University Press, 2007).

7   **ravens that were cheated refused:** J. J. A. Müller et al., "Ravens Remember the Nature of a Single Reciprocal Interaction Sequence over 2 Days and Even After a Month," *Animal Behaviour* 128 (2017): 69–78.

7   **Raven could not shake his curiosity:** National Storytelling Network, "Raven and the Whale," as retold by Laura Simms in 2001, https://storynet .org/raven-and-the-whale/#:~:text=an%20Inuit%20(Eskimo)%20story %2C,everything%20there%20was%20to%20know.

7   **raven populations for the past fifty years:** Y. Nihei and H. Higuchi, *When and Where Did Crows Learn to Use Automobiles as Nutcrackers?* (Tohoku University, 2002).

7   **the souls of ancient kings:** R. Graves, *The Greek Myths* (Penguin, 2012), 28.

9   **animals like the Superb parrot:** "Scout Camp, Dob, Dob, Dobs on Rogue Alligator," *Sydney Morning Herald,* December 31, 2008, https://www.smh.com.au/environment/scout-camp-dob-dob-dobs-on-rogue-alligator-20081231-gdt86m.html.

9   **world's second-most powerful bite:** S. Wroe et al., "Bite Club: Comparative Bite Force in Big Biting Mammals and the Prediction of Predatory Behaviour in Fossil Taxa," *Proceedings of the Royal Society B: Biological Sciences* 272, no. 1563 (2005): 619–25.

Part I: The Neighborhood Ark

14   **more biodiverse than the wild areas:** C. D. Ives et al., "Cities Are Hotspots for Threatened Species," *Global Ecology and Biogeography* 25, no. 1 (2016): 117–26.

14   **Cities *are* nature:** S. T. Pickett et al., "Shifting Forward: Urban Ecology in Perspective," *Ambio* 53 (2024): 890–97.

14   **shores of oceans, lakes, and rivers:** P. A. Todd et al., "Towards an Urban Marine Ecology: Characterizing the Drivers, Patterns and Processes of Marine Ecosystems in Coastal Cities," *Oikos* 128, no. 9 (2019): 1215–42.

15   **at-risk animals and plants:** R. I. McDonald et al., "The Implications of Current and Future Urbanization for Global Protected Areas and Biodiversity Conservation," *Biological Conservation* 141, no. 6 (2008): 1695–703.

Chapter 1: Puzzles

21   **spewing clouds of choking exhaust:** K. Jiang, "Toronto Traffic Is Among the Slowest in the World, New Report Suggests," *Toronto Star,* January 11, 2024, https://www.toronto.com/life/beauty-and-fashion/toronto-traffic-is-among-the-slowest-in-the-world-new-report-suggests/article_ef83c191-868f-5c58-b0a9-a7d0844f2db2.html.

25   **put Toronto's raccoon population at:** "Options for Mitigating Human-Wildlife Conflict in Toronto," in *City of Toronto Municipal Licensing and Standards* (City of Toronto, 2015).

25   **one hundred raccoons per square kilometer:** G. Salamone, "Raccoon-Raided Toronto Tries Coping with Resistant Garbage Bins and Coexisting Strategy," *Daily News,* October 9, 2018.

26   **National Wildlife Research Center:** L. A. Stanton et al., "Variation in Reversal Learning by Three Generalist Mesocarnivores," *Animal Cognition* 24, no. 3 (2021): 555–68.

30   **most widespread urban mammal:** C. G. Himsworth et al., "Rats, Cities, People, and Pathogens: A Systematic Review and Narrative Synthesis of Literature Regarding the Ecology of Rat-Associated Zoonoses in Urban Centers," *Vector-Borne and Zoonotic Diseases* 13, no. 6 (2013): 349–59.

31   **gave the rodents a lifeline:** J. Munshi-South et al., "The Evolutionary History of Wild and Domestic Brown Rats (*Rattus norvegicus*)," *Science* 385, no. 6715 (2024): 1292–97.

31   **rats culled from shipwrecks:** E. Guiry et al., "The Ratting of North America: A 350-Year Retrospective on Rattus Species Compositions and Competition," *Science Advances* 10, no. 14 (2024): eadm6755.

32   **NYC rat adaptations come in:** A. Harpak et al., "Genetic Adaptation in New York City Rats," *Genome Biology and Evolution* 13, no. 1 (2021): evaa247.

32   **a mutation in *CACNA1C*:** Harpak et al., "Genetic Adaptation in New York City Rats."

33   **autism, schizophrenia, depression, and:** A. L. Moon et al., "CACNA1C: Association with Psychiatric Disorders, Behavior, and Neurogenesis," *Schizophrenia Bulletin* 44, no. 5 (2018): 958–65.

33   **When they do socialize:** T. M. Kisko et al., "Cacna1c Haploinsufficiency Leads to Pro-Social 50-kHz Ultrasonic Communication Deficits in Rats," *Disease Models & Mechanisms* 11, no. 6 (2018): dmm034116.

33   **of a hard snowy city:** E. E. Puckett et al., "Variation in Brown Rat Cranial Shape Shows Directional Selection over 120 Years in New York City," *Ecology and Evolution* 10, no. 11 (2020): 4739–48.

33   **like plants and scavenged flesh:** V. Tapaltsyan et al., "Continuously Growing Rodent Molars Result from a Predictable Quantitative Evolutionary Change over 50 Million Years," *Cell Reports* 11, no. 5 (2015): 673–80.

34   **evolving in reverse to again:** Puckett et al., "Variation in Brown Rat Cranial Shape."

34   **uptown and downtown rats:** M. Combs et al., "Spatial Population Genomics of the Brown Rat (*Rattus norvegicus*) in New York City," *Molecular Ecology* 27, no. 1 (2018): 83–98.

37   **as far as thirty miles:** R. Rosatte et al., "Density, Movements, and Survival of Raccoons in Ontario, Canada: Implications for Disease Spread and Management," *Journal of Mammalogy* 91, no. 1 (2010): 122–35.

37   **more than three city blocks:** S. E. MacDonald and S. Ritvo, "Comparative Cognition Outside the Laboratory," *Comparative Cognition & Behavior Reviews* 11 (2016): 49–62.

37   **raccoons in its home range:** MacDonald and Ritvo, "Comparative Cognition Outside the Laboratory."

38   **"There's not a raccoon":** "Racoon-Resistant Green Bins Latest Weapon in Toronto's Compost Battle," CBC News, April 25, 2016, https://www.cbc.ca/news/canada/toronto/raccoon-green-bin-toronto-1.3552276.

40   **"the problem posed by the raccoon":** A. N. Iwaniuk and I. Q. Whishaw, "How Skilled Are the Skilled Limb Movements of the Raccoon (*Procyon lotor*)?" *Behavioural Brain Research* 99, no. 1 (1999): 35–44.

42   **devoted to their paws:** W. Welker and S. Seidenstein, "Somatic Sensory Representation in the Cerebral Cortex of the Racoon (*Procyon lotor*)," *Journal of Comparative Neurology* 111, no. 3 (1959): 469–501.

42   **closer to 15 percent:** Welker and Seidenstein, "Somatic Sensory
      Representation."

42   **receiving information from our hands:** A. R. Sobinov and S. J. Bensmaia,
      "The Neural Mechanisms of Manual Dexterity," *Nature Reviews
      Neuroscience* 22, no. 12 (2021): 741–57.

44   **rural parts of Ontario:** S. E. MacDonald and S. Ritvo, "Comparative
      Cognition Outside the Laboratory," *Comparative Cognition & Behavior
      Reviews* 11 (2016): 49–61.

45   **"support the tantalizing possibility":** MacDonald and Ritvo, "Comparative
      Cognition Outside."

45   **developed four theories about how:** L. Johnson-Ulrich et al., "Natural
      Conditions and Adaptive Functions of Problem-Solving in the Carnivora,"
      Current Opinion in Behavioral Sciences 44 (2022): 101111.

Chapter 2: The Tower

50   **more than eight hundred by the end:** R. Garcia, "Residential Foreclosures in
      the City of Buffalo, 1990-2000," *Federal Reserve Bank of New York* 2003: 1–74.

50   **twenty thousand vacant homes:** V. K. Carr, "Buffalo: Reinvention in a
      Shrinking City" (recent work, University of California, Berkeley, Graduate
      School of Journalism, 2012), https://escholarship.org/uc/item/43g758k5.

52   **end of the dinosaur age:** C. C. Voigt and T. Kingston, *Bats in the
      Anthropocene: Conservation of Bats in a Changing World* (Springer Nature,
      2016), 3.

52   **fragility of their bones:** E. E. Brown et al., "Quantifying the Completeness
      of the Bat Fossil Record," *Palaeontology* 62, no. 5 (2019): 757–76.

54   **building blocks, lie empty:** J. D. Epstein, "Downtown Office Market
      Struggles to Gain Its Footing," *Buffalo News*, February 15, 2024, https://
      buffalonews.com/news/local/business/downtown-buffalo-office-market
      -struggles-to-gain-its-footing/article_52b98736-caa3-11ee-a8ca-0b3d33
      bdc9ad.html.

55   **30 degrees Fahrenheit warmer:** A. N. Ahmed et al., "The Urban Heat Island
      Effect: A Review on Predictive Approaches Using Artificial Intelligence
      Models," *City and Environment Interactions* 28 (2025): 100234.

55   **eight hundred heartbeats per minute:** S. E. Currie et al., "Heart Rate as a
      Predictor of Metabolic Rate in Heterothermic Bats," *Journal of Experimental
      Biology* 217, pt. 9 (2014): 1519–24.

56   **more abundant in city centers:** G. Lesinski et al., "Foraging Areas and
      Relative Density of Bats (Chiroptera) in Differently Human Transformed
      Landscapes," *Zeitschrift fur Saugetierkunde* 65, no. 3 (2000): 129–37.

58   **solitary bees to lay their eggs:** J. S. MacIvor and L. Packer, "The Bees Among
      Us: Modelling Occupancy of Solitary Bees," *PLoS One* 11, no. 12 (2016):
      e0164764.

63   **evidence of fox synanthropy:** P. Wojtal et al., "The Scene of Spectacular

Feasts: Animal Remains from Pavlov I South-east, the Czech Republic," *Quaternary International* 252 (2012): 122–41.

63   **expansion of suburban communities trapped foxes:** "When and How Did Foxes Come to Live in Our Towns and Cities?" Wildlife Online, accessed March 10, 2025.

64   **people's brakes failed:** S. Quadri, "South London Residents 'Fox-Proof' Cars After Animals Chew Through Wires," *The Standard,* January 31, 2022.

64   **a cat-killing spree:** B. Perrigo, "It Took 3 Years for Scotland Yard to Conclude Foxes Were Behind an Alleged Cat Killing Spree in London," *Time,* September 20, 2018.

64   **only twenty-three reports of foxes biting:** B. Bridge and S. Harris, "Do Urban Red Foxes Attack People? An Exploratory Study and Review of Incidents in Britain," *Human-Wildlife Interactions* 14, no. 2 (2020): 151–65.

66   **190 vocalizations per second:** J. A. Thomas et al., *Echolocation in Bats and Dolphins* (University of Chicago Press, 2004), 136.

67   **"propagated oscillation":** American National Standards Institute, *American National Standard: Acoustical Terminology* (Acoustical Society of America, 2006), 1.

70   **the wild as "urban avoiders":** M. Schoeman, "Light Pollution at Stadiums Favors Urban Exploiter Bats," *Animal Conservation* 19, no. 2 (2016): 120–30; S. Kark et al., "Living in the City: Can Anyone Become an 'Urban Exploiter'?" *Journal of Biogeography* 34, no. 4 (2007): 638–51.

70   **since the late nineteenth century:** P. G. Newman and G. S. Rozycki, "The History of Ultrasound," *Surgical Clinics of North America* 1998; 78, no. 2: 179–95.

71   **need unimpeded communication to hunt:** M. Grimshaw-Aagaard and B. Bemman, "Ultrasonics and Urban Greening: An Exploratory Study on Ultrasound Presence in Urban Spaces," *Personal and Ubiquitous Computing* 28, no. 5 (2024): 677–92.

71   **as much as 70 percent:** J. P. Bunkley et al., "Anthropogenic Noise Alters Bat Activity Levels and Echolocation Calls," *Global Ecology and Conservation* 3 (2015): 62–71.

71   **their arsenal to catch prey:** D. Russo and L. Ancillotto, "Sensitivity of Bats to Urbanization: A Review," *Mammalian Biology* 80, no. 3 (2015): 205–12.

72   **pouch to catch their prey:** E. Amichai and C. Korine, "Kuhl's Pipistrelle *Pipistrellus kuhlii* (Kuhl, 1817)," *Handbook of the Mammals of Europe* (2020): 1–19.

72   **the range of human hearing:** D. Russo and G. Jones, "The Social Calls of Kuhl's Pipistrelles *Pipistrellus kuhlii* (Kuhl, 1819): Structure and Variation (Chiroptera: Vespertilionidae)," *Journal of Zoology* 249, no. 4 (1999): 469–93.

73   **mentions the #mprraccoon:** M. Haag and C. Caron, "Masked Fugitive, at Ease in the Dark, Scales a Skyscraper in Minnesota," *New York Times,* June 13, 2018, A17.

75   **The complex local dialect:** N. Starik and T. Göttert, "Bats Adjust

Echolocation and Social Call Design as a Response to Urban Environments," *Frontiers in Ecology and Evolution* 10 (2022): 939408.

76   **"honest signal of body size":** S. J. Puechmaille et al., "Female Mate Choice Can Drive the Evolution of High Frequency Echolocation in Bats: A Case Study with *Rhinolophus mehelyi*," *PLoS One* 9, no. 7 (2014): e103452.

77   **mid-December to mid-January:** M. Weinberg et al., "Urban Fruit Bats Give Birth Earlier in the Season Compared to Rural Fruit Bats," *BMC Biology* 23, no. 31 (2025), https://bmcbiol.biomedcentral.com/articles/10.1186/s12915 -025-02124-y.

Chapter 3: The Soft Intelligence

83   **dozens of other dangerous substances:** "Water Quality Atlas, Olympia, WA," Washington State Department of Ecology, https://apps.ecology.wa.gov /waterqualityatlas/wqa/map.

85   **But most of the world's largest:** B. Neumann et al., "Future Coastal Population Growth and Exposure to Sea-Level Rise and Coastal Flooding—A Global Assessment," *PLoS One* 10, no. 3 (2015): e0118571.

86   **and bits of old metal:** W. Davis III and A. G. Murphy, "Plastic in Surface Waters of the Inside Passage and Beaches of the Salish Sea in Washington State," *Marine Pollution Bulletin* 97, nos. 1–2 (2015): 169–77.

86   **In just a single square meter:** F. Eshom-Arzadon, "Concentration of Microplastics in Beach Sediments Surrounding Seattle, Washington, in the Puget Sound Estuary" (master's thesis, University of Washington, 2017).

91   **their lives attached to rocks:** L. Airoldi et al., "Corridors for Aliens but Not for Natives: Effects of Marine Urban Sprawl at a Regional Scale," *Diversity and Distributions* 21, no. 7 (2015): 755–68.

93   **process called chromatic aberration:** A. L. Stubbs and C. W. Stubbs, "Spectral Discrimination in Color Blind Animals Via Chromatic Aberration and Pupil Shape," *Proceedings of the National Academy of Sciences* 113, no. 29 (2016): 8206–11.

94   **and its level of biodiversity:** M. Leong et al., "Biodiversity and Socioeconomics in the City: A Review of the Luxury Effect," *Biology Letters* 14, no. 5 (2018): 20180082.

94   **where the wealth is:** S. B. Magle et al., "Urban Mesopredator Distribution: Examining the Relative Effects of Landscape and Socioeconomic Factors," *Animal Conservation* 19, no. 2 (2016): 163–75.

96   **are filled with GPOs:** A. Acevedo-Gutiérrez et al., "First Diet Description of the Harbor Seal (*Phoca vitulina*) in the Northwest Corner of the Olympic Peninsula, Washington State," *Northwestern Naturalist* 104, no. 3 (2023): 253–65; M. K. Trzcinski et al., "DNA Analysis of Scats Reveals Spatial and Temporal Structure in the Diversity of Harbour Seal Diet from Local Haulouts to Oceanographic Bioregions," *Marine Ecology Progress Series* 743 (2024): 113–38.

96 **Pinnipeds are fond:** J. Quezada, " 'We Want It for Humans Only': San Diego Group Pushes to Get Sea Lions Out of La Jolla Cove," NBC 7 San Diego, July 11, 2024, https://www.nbcsandiego.com/news/local/la-jolla-cove-sea -lion-management-plan/3564143/.

103 **GPOs produce up to four:** S. Larson et al., "Multiple Paternity and Preliminary Population Genetics of Giant Pacific Octopuses, *Enteroctopus dofleini,* in Oregon, Washington and the Southeast Coast of Vancouver Island, BC," *Diversity* 7, no. 2 (2015): 195–205.

Chapter 4: Grave Architecture

105 **major city in North America:** S. A. Poessel et al., "Environmental Factors Influencing the Occurrence of Coyotes and Conflicts in Urban Areas," *Landscape and Urban Planning* 157 (2017): 259–69.

105 **fifteen-fold during the 1990s:** S. Gehrt, *Urban Coyote Ecology and Management* (Cook County, Illinois, Coyote Project, 2011), 4.

106 **The biologists, who collared:** W. F. Andelt and B. R. Mahan, "Behavior of an Urban Coyote," *American Midland Naturalist* 103, no. 2 (1980): 399–400.

107 **an astounding 19 percent:** E. L. Glaeser and J. M. Shapiro, *City Growth and the 2000 Census: Which Places Grew, and Why* (Brookings Institution, Center on Urban and Metropolitan Policy, 2001), 1, https://www2.lawrence .edu/fast/finklerm/whygrowth.pdf.

107 **plains 1 million years ago:** K. Myers, "Coyote-A Wily, New Emblem for the Anthropocene? Why One of the Most Reviled Predators in North America Could Change Conservation Forever" (master's thesis, Georgia State University, 2021).

108 **points, like in the 1950s:** H. M. Fener et al., "Chronology of Range Expansion of the Coyote, *Canis latrans,* in New York," *Canadian Field- Naturalist* 119, no. 1 (2005): 1–5.

108 **their presence was overlooked:** Fener et al., "Chronology of Range Expansion."

108 **Between 1630 and 1930:** X. Li et al., "Four-Century History of Land Transformation by Humans in the United States (1630–2020): Annual and 1km Grid Data for the History of Land Changes (HISLAND-US)," *Earth System Science Data* 15, no. 2 (2023): 1005–35.

114 **Instead, their paths avoided:** C. E. Wilkinson et al., "Environmental Health and Societal Wealth Predict Movement Patterns of an Urban Carnivore," *Ecology Letters* 28, no. 2 (2025): e70088.

116 **roughly forty-four thousand insect species:** K. A. Triantis et al., "Extinction Debt on Oceanic Islands," *Ecography* 2010; 33, no. 2: 285–94.

117 **the past six hundred years:** Wilkinson et al., "Environmental Health and Societal Wealth."

117 **One broad rule of thumb:** K. A. Triantis et al., "Extinction Debt on Oceanic Islands," 285–94.

117   **be possible given its size:** J. M. Diamond, "Biogeographic Kinetics: Estimation of Relaxation Times for Avifaunas of Southwest Pacific Islands," *Proceedings of the National Academy of Sciences* 69, no. 11 (1972): 3199–203.

118   **What's more, the inbred cats:** A. A. Huffmeyer et al., "First Reproductive Signs of Inbreeding Depression in Southern California Male Mountain Lions (*Puma concolor*)," *Theriogenology* 177 (2022): 157–64.

120   *Have you witnessed your pet*: C. A. Niesner et al., "The Coyote in the Cloud," *Environment and Planning E: Nature and Space* 7, no. 3 (2024): 1054–75.

120   **wily as Coyote himself:** K. B. Judson, *Myths and Legends of California and the Old Southwest* (University of Nebraska Press, 1994), 28.

121   **worshipped deity in North America:** Myers, "Coyote-A Wily."

121   **entirely mimicked a city environment:** S. A. Poessel et al., "Influence of Habitat Structure and Food on Patch Choice of Captive Coyotes," *Applied Animal Behaviour Science* 157 (2014): 127–36.

122   **This explains, says Schell:** C. J. Schell et al., "Parental Habituation to Human Disturbance Over Time Reduces Fear of Humans in Coyote Offspring," *Ecology and Evolution* 8, no. 24 (2018): 12965–80.

122   **only as a last resort:** S. W. Breck et al., "Evaluating Lethal and Nonlethal Management Options for Urban Coyotes," *Human-Wildlife Interactions* 11, no. 2 (2017): 133–45.

122   **or so are killed annually:** "Media Release: 2023 Wildlife Services Report: Taxpayer Dollars Used to Kill Over 70,000 Wild Carnivores," Project Coyote, April 15, 2024, https://projectcoyote.org/media-release-2023-wildlife -services-report-taxpayer-dollars-used-to-kill-over-70000-wild-carnivores/.

122   **Like many synanthropes, coyotes:** K. Atkinson and D. Shackleton, "Coyote, Canis Latrans, Ecology in a Rural-Urban Environment," *Canadian Field-Naturalist* 105, no. 1 (1991): 49–54; S. Grant et al., *Assessment of Human–Coyote Conflicts: City and County of Broomfield, Colorado* (USDA National Wildlife Research Center, 2011), https:// digitalcommons.unl.edu/cgi/viewcontent.cgi?article=2216&context =icwdm_usdanwrc; M. I. Grinder and P. R. Krausman, "Home Range, Habitat Use, and Nocturnal Activity of Coyotes in an Urban Environment," *Journal of Wildlife Management* 65, no. 4 (2001): 887–98; and T. Quinn, "Coyote (*Canis latrans*) Food Habits in Three Urban Habitat Types of Western Washington," *Northwest Science* 71, no. 1 (1997): 1–5.

123   **"ways to survive in cities":** Quotes from: J. Davidow, "Seeing Himself in the Science," *University of Washington Magazine,* December 2019; C. Larson, "Calling Out Racism: Black Scientists Say They Face Discrimination While Doing Fieldwork," *Chicago Sun-Times,* September 15, 2020.

126   **"It was a safe place":** Schell quotes from: G. Kell, "After the LA Wildfires, Stories Emerge at UC Berkeley of Great Loss, Support and Strength," *UC Berkeley News,* February 7, 2025, https://news.berkeley.edu/2025/02/07/after -the-la-wildfires-stories-emerge-at-uc-berkeley-of-great-loss-support-and -strength/.

127 **"Becoming a statistic after":** Schell quote from: A. Mahoney, "Escaping to Altadena Only to Flee Flames," *In These Times,* February 10, 2025, https://inthesetimes.com/article/la-wildfires-impact-altadena-black-community.

127 **Across town, on Palisades Drive:** R. Straker, "California Wildfires Hurting Wildlife: Burned Coyote Wanders Pacific Palisades Street," Weather Channel, January 27, 2025, https://www.yahoo.com/news/california-wildfires-hurting-wildlife-burned-173146740.html.

129 **a better chance of survival:** B. A. Robertson and R. L. Hutto, "A Framework for Understanding Ecological Traps and an Evaluation of Existing Evidence," *Ecology* 87, no. 5 (2006): 1075–85.

Part II: Damnatio ad Bestias

134 **known as *insula*:** M. MacKinnon, "Pack Animals, Pets, Pests, and Other Non-Human Beings," in *The Cambridge Companion to Ancient Rome,* ed. P. Erdkamp (Cambridge University Press, 2013), 110.

134 **replication of identical habitats:** C. S. Elton, *The Ecology of Invasions by Animals and Plants* (University of Chicago Press, 2000), 53.

134 **becoming increasingly homogenous:** M. Stuhlmacher et al., "Are Global Cities Homogenizing? An Assessment of Urban Form and Heat Island Implications," *Cities* 126 (2022): 103705.

136 **"biotic homogenization":** M. L. McKinney, "Urbanization as a Major Cause of Biotic Homogenization," *Biological Conservation* 127, no. 3 (2006): 247–60.

Chapter 5: Neophytes

139 **rhizomes weigh up to four hundred pounds:** R. J. Blaustein, "Kudzu's Invasion into Southern United States Life and Culture," in *The Great Reshuffling: Human Dimensions of Invasive Species,* ed. J. A. McNeeley (IUCN, 2001), 56.

140 **The perambulating vine:** T. R. Kartzinel et al., "Heterogeneity of Clonal Patterns Among Patches of Kudzu, *Pueraria montana* var. lobata, an Invasive Plant," *Annals of Botany* 116, no. 5 (2015): 739–50.

141 **three hundred species at risk of extinction:** "Key Facts About Rabbit Biocontrol in Australia: A Brief History of Rabbits in Australia," Centre for Invasive Species Solutions, 2025; Centre for Invasive Species Solutions, "European Rabbits: Key Facts About Rabbit Biocontrol in Australia," PestSmart, 2015, https://pestsmart.org.au/toolkit-resource/key-facts-about-rabbit-biocontrol-in-australia/#:~:text=Impacts%20of%20rabbits%20in%20Australia,less%20effective%20in%20cooler%20climates.

142 **your lawn, with sales:** "Seed, Kentucky Blue Grass, for Sowing," Observatory of Economic Complexity, https://oec.world/en/profile/hs/seed-kentucky-blue-grass-for-sowing.

142   **contemporary weeds at a hunter-gatherer site:** A. Snir et al., "The Origin of
       Cultivation and Proto-Weeds, Long Before Neolithic Farming," *PLoS One* 10,
       no. 7 (2015): e0131422.

142   **first-ever census of floral synanthropes:** I. Kowarik and P. Pyšek, "The
       First Steps Towards Unifying Concepts in Invasion Ecology Were
       Made One Hundred Years Ago: Revisiting the Work of the Swiss Botan-
       ist Albert Thellung," *Diversity and Distributions* 18, no. 12 (2012):
       1243–52.

144   **distributed to farmers across the South:** I. N. Forseth and A. F. Innis,
       "Kudzu (*Pueraria montana*): History, Physiology, and Ecology Combine to
       Make a Major Ecosystem Threat," *Critical Reviews in Plant Sciences* 23, no. 5
       (2004): 401–13.

144   **"healing touch of the miracle vine":** B. Finch, "The True Story of Kudzu, the
       Vine That Never Truly Ate the South," *Smithsonian,* September 2015, https://
       www.smithsonianmag.com/science-nature/true-story-kudzu-vine-ate-south
       -180956325/.

145   **should have sounded:** "To the Spades, Men!" *The Birmingham News,*
       August 14, 1959.

146   **"this sucker could go down":** D. M. Herszenhorn et al., "Talks Implode
       During a Day of Chaos; Fate of Bailout Plan Remains Unresolved," *New York
       Times,* September 26, 2008, https://www.nytimes.com/2008/09/26/business
       /26bailout.html?hp.

155   **compared to a plant that needs a mate:** P.-O. Cheptou, "Clarifying Baker's
       Law," *Annals of Botany* 109, no. 3 (2011): 633–41.

155   **ride out the initial period of scarcity:** Forseth and Innis, "Kudzu (*Pueraria
       montana*)."

156   **plant's genetic diversity is sky-high:** R. A. Pappert et al., "Genetic Variation
       in *Pueraria lobata* (Fabaceae), an Introduced, Clonal, Invasive Plant of the
       Southeastern United States," *American Journal of Botany* 87, no. 9 (2000):
       1240–45.

156   **creeping ever forward in all directions:** Kartzinel et al., "Heterogeneity of
       Clonal Patterns."

161   **It thrives in every:** I checked, and it's not growing in the underpass beside
       the Nick, the legendary punk bar on Birmingham's Southside, which you
       should probably not walk to alone at night. But if you go inside, there is a
       young woman who sometimes sits at the bar who will sell you a knife,
       which is a fine consolation prize.

162   **causes the plant to die out:** H. Tsugawa et al., "Influence of Shade
       Treatment on Leaf and Branch Emergence, and Dry Matter Production of
       Kudzu Vine Seedlings (*Pueraria lobata Ohwi*)," 神戸大学農学部研究報告
       (Research Report of the Faculty of Agriculture, Kobe University) 16, no. 2
       (1985): 359–67, https://da.lib.kobe-u.ac.jp/da/kernel/00225553/00225553.pdf.

163   **be spreading at a rate:** Forseth and Innis, "Kudzu (*Pueraria montana*)."

163   **its spread across the continent:** Finch, "The True Story of Kudzu."

Chapter 6: The New Animism

168 **born with only one set of teeth:** S. Pappas, "Why Are Killer Whales Ripping Livers Out of Their Shark Prey?" *Scientific American*, April 11, 2023, https://www.scientificamerican.com/article/why-do-killer-whales-rip-out-shark-livers/.

168 **newfound toys die of exhaustion:** J. Felton, "In 1987, Orcas Had a Fashion of Wearing a Dead Salmon as a Hat," IFLScience, June 27, 2023, https://www.iflscience.com/in-1987-orcas-had-a-fashion-of-wearing-a-dead-salmon-as-a-hat-69542.

173 **temporary land bridge at the southern tip of the Red Sea:** G. H. Kopp et al., "Out of Africa, but How and When? The Case of Hamadryas Baboons (*Papio hamadryas*)," *Journal of Human Evolution* 76 (2014): 154–64.

174 **travel as far as thirteen kilometers per day:** A. Cherion, "Hamadryas Baboon, *Papio hamadryas*," New England Primate Conservancy, October 2022, https://neprimateconservancy.org/hamadryas-baboon/#:~:text=They%20can%20travel%202-8,recorded%20mostly%20in%20captive%20individuals.

175 **in 1950, less than 10 percent:** "Population Development in Saudi Arabia," WorldData.info, https://www.worlddata.info/asia/saudi-arabia/population growth.php.

175 **If a baboon raided a farm, it was shot and killed:** S. Biquand et al., "The Distribution of *Papio hamadryas* in Saudi Arabia: Ecological Correlates and Human Influence," *International Journal of Primatology* 13, no. 3 (1992): 223–43.

178 **ate them on the car hoods:** R. Harrison, "Traditional Enemies Live in Harmony in Talif," *Arab News*, January 1, 2012, https://www.arabnews.com/node/403441#:~:text=A%20simpler%20explanation%20might%20lie,refurbishment%20of%20the%20escarpment%20road.

179 **fewer than three hundred baboons:** Biquand et al., "The Distribution of *Papio hamadryas*."

181 **killed shortly after it is born:** A. L. Amann et al., "Contexts and Consequences of Takeovers in Hamadryas Baboons: Female Parity, Reproductive State, and Observational Evidence of Pregnancy Loss," *American Journal of Primatology* 79, no. 7 (2017): e22649.

181 **same brown color as females:** H. Kummer, "A Male Dominated Society: The Hamadryas Baboon of Cone Rock, Ethiopia," in *The Encyclopedia of Mammals*, ed. D. MacDonald (Oxford University Press, 2001), 376.

182 **around 400,000 years ago:** D. E. Wildman et al., "Mitochondrial Evidence for the Origin of Hamadryas Baboons," *Molecular Phylogenetics and Evolution* 32, no. 1 (2004): 287–96.

182 **"wandering females":** "National Plan to Solve the Baboon Problem by 2026," *Wildlife Magazine*, December 4, 2024, https://ncw.gov.sa/magazine/en/article/national-plan-to-solve-the-baboon-problem-by-2026/.

182   **a calming ripple effect:** A. Mori et al., "A Study on the Social Structure and Dispersal Patterns of Hamadryas Baboons Living in a Commensal Group at Taif, Saudi Arabia," *Primates* 48, no. 3 (2007): 179–89.

184   **a battered feral house cat:** Feral animals, incidentally, are previously domesticated species that have broken the psychological chains binding them to humans and relearned how to survive in the wild. Pigeons, dogs, cats, and (in al-Baha, at least) donkeys are examples of animals that humans have abandoned to nature's care.

187   **risk of being killed or seriously injured:** E. van der Meer et al., "Dangerous Game: Preferential Predation on Baboons by African Wild Dogs in Mana Pools National Park, Zimbabwe," *Behaviour* 156, no. 1 (2019): 37–58.

187   **"he ceases to fear its touch":** Quoted from: E. Canetti, *Crowds and Power* (Macmillan, 1984), 10.

191   **forcing the submission of feral dogs:** Craig and Pamelia, "Do Baboons Keep Dogs as Pets?" *News, Notes and Photos from the Field* (blog), Naturalist's Notebook, February 17, 2015, https://www.thenaturalistsnotebook.com/our -blog/do-baboons-keep-dogs-as-pets.

192   **maximum range of hamadryas baboons:** M. C. Henriquez et al., "Home Range, Sleeping Site Use, and Band Fissioning in Hamadryas Baboons: Improved Estimates Using GPS Collars," *American Journal of Primatology* 83, no. 5 (2021): e23248.

Chapter 7: Deep Time

194   **HERVs:** D. J. Griffiths, "Endogenous Retroviruses in the Human Genome Sequence," *Genome Biology* 2, no. 6 (2001): reviews1017.

194   **measles, Ebola, avian flu, and coronaviruses:** S. S. Badarinarayan and D. Sauter, "Switching Sides: How Endogenous Retroviruses Protect Us from Viral Infections," *Journal of Virology* 95, no. 12 (2021): e02299–320.

196   **scat placed on their torsos:** E. L. Jenkins, "Mice, Scats and Burials: Unusual Concentrations of Microfauna Found in Human Burials at the Neolithic Site of Çatalhöyük, Central Anatolia," *Journal of Social Archaeology* 12, no. 3 (2012): 380–403.

196   **small synanthropic rodents:** M. Krajcarz et al., "Ancestors of Domestic Cats in Neolithic Central Europe: Isotopic Evidence of a Synanthropic Diet," *Proceedings of the National Academy of Sciences* 117, no. 30 (2020): 17710–19.

196   **became synanthropic more than twelve thousand years ago:** T. Cucchi et al., "Tracking the Near Eastern Origins and European Dispersal of the Western House Mouse," *Scientific Reports* 10, no. 1 (2020): 8276.

196   **bones of a human infant:** I. Hershkovitz et al., "Detection and Molecular Characterization of 9000-Year-Old Mycobacterium Tuberculosis from a Neolithic Settlement in the Eastern Mediterranean," *PLoS One* 3, no. 10 (2008): e3426.

196   **undercooked poultry or eggs:** F. M. Key et al., "Emergence of Human-Adapted Salmonella Enterica Is Linked to the Neolithization Process," *Nature Ecology & Evolution* 4, no. 3 (2020): 324–33.

197   *Yersinia pestis*: N. Bergfeldt et al., "Identification of Microbial Pathogens in Neolithic Scandinavian Humans," *Scientific Reports* 14, no. 1 (2024): 5630.

197   **dating back to Neolithic India:** H. Yu et al., "Palaeogenomic Analysis of Black Rat (*Rattus rattus*) Reveals Multiple European Introductions Associated with Human Economic History," *Nature Communications* 13, no. 1 (2022): 2399.

197   **mpox:** M. Curaudeau et al., "Identifying the Most Probable Mammal Reservoir Hosts for Monkeypox Virus Based on Ecological Niche Comparisons," *Viruses* 15, no. 3 (2023): 727.

197   **hepatitis B virus:** J. F. Drexler et al., "Bats Carry Pathogenic Hepadnaviruses Antigenically Related to Hepatitis B Virus and Capable of Infecting Human Hepatocytes," *Proceedings of the National Academy of Sciences* 110, no. 40 (2013): 16151–56.

197   **rinderpest around 6000 B.C.E.:** A. Düx et al., "Measles Virus and Rinderpest Virus Divergence Dated to the Sixth Century BCE," *Science* 368, no. 6497 (2020): 1367–70.

197   **the most infectious virus:** A. A. Guzmán-Solís et al., "Glimpse into the Past: What Ancient Viral Genomes Reveal About Human History," *Annual Review of Virology* 10, no. 1 (2023): 49–75.

198   **enough of us to adapt to and kill:** L. M. Van Blerkom, "Role of Viruses in Human Evolution," *American Journal of Physical Anthropology* 122, no. S37 (2003): 14–46.

198   **"the first epidemiologic transition":** G. F. Armelagos, "Emerging Disease in the Third Epidemiological Transition," in *The Changing Face of Disease: Implications for Society,* ed. C. G. N. Mascie-Taylor, J. Peters, and S. T. McGarvey (CRC Press, 2004), 12.

201   **a staggering 50 percent:** Public Health Agency of Canada, *Human Emerging Respiratory Pathogens Bulletin* 94 (Centre for Emerging and Respiratory Infections and Pandemic Preparedness; Public Health Agency of Canada, October 2024), 3.

202   **the entirety of the twentieth century:** R. M. Meganck and R. S. Baric, "Developing Therapeutic Approaches for Twenty-First-Century Emerging Infectious Viral Diseases," *Nature Medicine* 27, no. 3 (2021): 401–10.

202   **It was the third, following quickly after SARS:** B. Ganesh et al., "Epidemiology and Pathobiology of SARS-CoV-2 (COVID-19) in Comparison with SARS, MERS: An Updated Overview of Current Knowledge and Future Perspectives," *Clinical Epidemiology and Global Health* 10 (2021): 100694.

203   **sickened schoolchildren in Haiti:** J. A. Lednicky et al., "Independent Infections of Porcine Deltacoronavirus Among Haitian Children," *Nature* 600, no. 7887 (2021): 133–37.

203 **exposed to wildlife in Malaysia:** A. N. Vlasova et al., "Novel Canine Coronavirus Isolated from a Hospitalized Patient with Pneumonia in East Malaysia," *Clinical Infectious Diseases* 74, no. 3 (2022): 446–54.

203 **norovirus:** R. M. Meganck and R. S. Baric, "Developing Therapeutic Approaches for Twenty-First-Century Emerging Infectious Viral Diseases," *Nature Medicine* 27, no. 3 (2021): 401–10.

204 **ending the day well-fed:** D. Thornton et al., "Hunting Associations of American Badgers (*Taxidea taxus*) and Coyotes (*Canis latrans*) Revealed by Camera Trapping," *Canadian Journal of Zoology* 96, no. 7 (2018): 769–73.

209 **group of SARS-like bat coronaviruses:** M. F. Boni et al., "Evolutionary Origins of the SARS-CoV-2 Sarbecovirus Lineage Responsible for the COVID-19 Pandemic," *Nature Microbiology* 5, no. 11 (2020): 1408–17.

209 **between 1881 and 1918:** B. F. Keele et al., "Chimpanzee Reservoirs of Pandemic and Nonpandemic HIV-1," *Science* 313, no. 5786 (2006): 523–26.

209 **rural southeast Cameroon sometime around 1920:** S. Gryseels et al., "A Near Full-Length HIV-1 Genome from 1966 Recovered from Formalin-Fixed Paraffin-Embedded Tissue," *Proceedings of the National Academy of Sciences* 117, no. 22 (2020): 12222–29.

209 **There are 10^{31}:** A. R. Mushegian, "Are There 10(31) Virus Particles on Earth, or More, or Fewer?" *Journal of Bacteriology* 202, no. 9 (2020).

210 **in the Huanan market in Wuhan, during November and December 2019:** J. E. Pekar et al., "The Molecular Epidemiology of Multiple Zoonotic Origins of SARS-CoV-2," *Science* 377, no. 6609 (2022): 960–66.

212 **"You can buy a Jet Ski":** "Mindful Solutionism," by Aesop Rock, produced by Aesop Rock, Rhymesayers, released as a single on September 14, 2023.

213 **infected across seventeen states:** "Current Situation: Bird Flu in Dairy Cows," CDC, July 7, 2025, https://www.cdc.gov/bird-flu/situation-summary/mammals.html.

213 **barn cats tested positive for the virus:** E. R. Burrough et al., "Highly Pathogenic Avian Influenza A (H5N1) Clade 2.3.4.4b Virus Infection in Domestic Dairy Cattle and Cats, United States, 2024," *Emerging Infectious Diseases* 30, no. 7 (2024): 1335.

213 **every time it replicates:** The exception are coronaviruses, which are unique among RNA viral families in having a proofreader, allowing them to have much more complex genomes (roughly thirty thousand nucleotide bases), but impeding them from evolving as quickly as other RNA viruses.

216 **strains that can infect humans:** R. D. Siegel, "Classification of Human Viruses," *Principles and Practice of Pediatric Infectious Diseases* (July 18, 2017): 1044–48.e1.

216 **twelve different group 2 coronaviruses:** D. R. Martinez et al., "Vaccine-Mediated Protection Against Merbecovirus and Sarbecovirus challenge in Mice," *Cell Reports* 42, no. 10 (2023): 113248; and D. Werb, "The U.S. Scientist at the Heart of COVID-19 Lab Leak Conspiracies Is Still Trying to Save the

World from the Next Pandemic," *Time*, July 11, 2023, https://time.com/6290193/covid-lab-leak-ralph-baric/.

217 **creating a blended vaccine:** E. P. Mazunina et al., "Trivalent mRNA Vaccine-Candidate Against Seasonal Flu with Cross-Specific Humoral Immune Response," *Frontiers in Immunology* 15 (2024): 1381508.

218 **eliminate pandemic threats:** World Health Organization, Food and Agriculture Organization of the United Nations, World Organisation for Animal Health, and United Nations Environment Programme, *One Health Joint Plan of Action (2022–2026): Working Together for the Health of Humans, Animals, Plants and the Environment* (World Health Organization, 2022).

Part III: The Synanthropocene

221 **become inhospitable to non-human life-forms:** M. F. J. Aronson et al., "A Global Analysis of the Impacts of Urbanization on Bird and Plant Diversity Reveals Key Anthropogenic Drivers," *Proceedings of the Royal Society B: Biological Sciences* 281, no. 1780 (2014): 20133330.

222 **parrots are content to eat trash:** Á. Luna et al., "Cities May Save Some Threatened Species but Not Their Ecological Functions," *PeerJ* 6 (2018): e4908.

Chapter 8: Embracing the Widow Makers

226 **"crisis disciplines":** M. E. Soulé, "What Is Conservation Biology? A New Synthetic Discipline Addresses the Dynamics and Problems of Perturbed Species, Communities, and Ecosystems," *BioScience* 35, no. 11 (1985): 727–34.

232 **"whoever prays to Her, Mother Bonbibi protects":** A. Sen and J. Mukherjee, "Bonbibi: A Religion of the Forest in the Sundarbans," *Arcadia* 22 (2020), 1.

233 **force their jaws shut:** A. N. Chowdhurym et al., "Ecopsychosocial Aspects of Human–Tiger Conflict: An Ethnographic Study of Tiger Widows of Sundarban Delta, India," *Environmental Health Insights* 10, no. 1 (2016): S24899.

233 **"cautiously o boatman":** P. Mukhopadhyay, "The Fisherman of the Forest: A Socio-Anthropological Study of the Landscape of the Deltaic Sundarbans in Bengal" (PhD diss., University of Delhi, 2021), https://shodhganga.inflibnet.ac.in/handle/10603/477257.

238 **"violating thereby the Forest Act":** D. Chowdhury, "Space, Identity, Territory: Marichjhapi Massacre, 1979," *International Journal of Human Rights* 15, no. 5 (2011): 664–82.

238 **dumped them there for the animals to feed on:** M. B. Dey, "The Forgotten Genocide: How the Namasudras Were Betrayed Twice," News18, January 30, 2025, https://www.news18.com/opinion/opinion-the-forgotten-genocide-how-the-namasudras-were-betrayed-twice-9207123.html.

238 **"tiger food":** A. Jalais, "Dwelling on Morichjhanpi: When Tigers Became 'Citizens,' Refugees 'Tiger-Food,'" *Economic and Political Weekly* 40, no. 17 (2005): 1757–62.

239 **burned corpses at the tigers' feet:** S. Montgomery, *Spell of the Tiger: The Man-Eaters of Sundarbans* (Chelsea Green, 2009), 14.

239 **two hundred meters of coastline:** X. Lu, "The World's Largest Mangrove Forest Is Shrinking," Earth.org, October 8, 2019, https://earth.org/the-worlds -largest-mangrove-forest-is-shrinking/.

240 **they can't swim more than:** M. A. Aziz, "Population Status, Threats, and Evolutionary Conservation Genetics of Bengal Tigers in the Sundarbans of Bangladesh" (PhD diss., University of Kent, 2017), https://kar.kent.ac .uk/61422/.

241 **a death knell for the population:** S. P. D. Riley et al., "Individual Behaviors Dominate the Dynamics of an Urban Mountain Lion Population Isolated by Roads," *Current Biology* 24, no. 17 (2014): 1989–94.

241 **predatory, territorial, or defensive:** C. D. Soulsbury and P. C. L. White, "Human–Wildlife Interactions in Urban Areas: A Review of Conflicts, Benefits and Opportunities," *Wildlife Research* 42, no. 7 (2015): 541–53.

241 **a disturbing linear relationship:** G. Yirga et al., "Peri-Urban Spotted Hyena (*Crocuta crocuta*) in Northern Ethiopia: Diet, Economic Impact, and Abundance," *European Journal of Wildlife Research* 57, no. 4 (2011): 759–65.

241 **necessary to keep themselves and their families alive:** G. Bombieri et al., "A Worldwide Perspective on Large Carnivore Attacks on Humans," *PLoS Biology* 21, no. 1 (2023): e3001946.

243 **take the place of their late husbands:** A. N. Chowdhurym et al., "Ecopsychosocial Aspects of Human–Tiger Conflict: An Ethnographic Study of Tiger Widows of Sundarban Delta, India," *Environmental Health Insights* 10, no. 1 (2016): S24899.

244 **scale measuring the stigma they experience:** Chowdhurym et al., "Ecopsychosocial Aspects of Human–Tiger Conflict."

244 **post-traumatic stress disorder:** Chowdhurym et al., "Ecopsychosocial Aspects of Human–Tiger Conflict."

244 **almost 80 percent were home to a tiger widow:** P. Mukhopadhyay, "Living in 'Limbo': Death in Everyday Sundarbans," in *Death and Events,* ed. I. R. Lamond and R. Dowson (Routledge, 2021), 113–29.

246 **between one and two dollars a day:** P. Sarkar et al., "Contribution of Mangrove Ecosystem Services to Local Livelihoods in the Indian Sundarbans," *Sustainability* 16, no. 16 (2024): 6804.

247 **"resties":** Megnaa Mehtta, "A 'License' to Kill in the Sundarbans," *India Forum,* November 10, 2021, https://www.theindiaforum.in/article/licence -kill-sundarbans.

247 **livelihoods of about 2 million people:** S. Chacraverti, *The Sundarbans Fishers: Coping in an Overly Stressed Mangrove Estuary* (International Collective in Support of Fishworkers, 2014), 34.

248 **women and girls from the Sundarbans:** P. D. Lobo, "Unmasking Dark Malady of Sex Trafficking in the Sunderbans [*sic*]," *Daily Guardian*, March 25, 2023.

Chapter 9: Listening to Animals

252   *oikouménē*: Literally, "the area inhabited by people."

254   **Two-thirds of them die after colliding with cars:** A. D. Rodewald and S. D. Gehrt, "Wildlife Population Dynamics in Urban Landscapes," in *Urban Wildlife Conservation: Theory and Practice,* ed. R. McCleery, C. Moorman, and M. Peterson (Springer, 2014), 117–47.

264   **baby salmon have flooded back:** A. C. Sawyer et al., "Seawall as Salmon Habitat: Eco-Engineering Improves the Distribution and Foraging of Juvenile Pacific Salmon," *Ecological Engineering* 151 (2020): 105856.

265   **the calmer the squirrels became:** T. S. Parker and C. H. Nilon, "Urban Landscape Characteristics Correlated with the Synurbization of Wildlife," *Landscape and Urban Planning* 106, no. 4 (2012): 316–25.

269   **"stepping stones":** A. J. Lynch, "Creating Effective Urban Greenways and Stepping-Stones: Four Critical Gaps in Habitat Connectivity Planning Research," *Journal of Planning Literature* 34, no. 2 (2019): 131–55.

271   **"supertrees":** L. Said-Moorhouse, "Solar-Powered 'Supertrees' Breathe Life into Singapore's Urban Oasis," CNN, June 10, 2012, https://www.cnn .com/2012/06/08/world/asia/singapore-supertrees-gardens-bay/index.html.

272   **true number could be six thousand or more:** M. Mehtta, "Counting Tigers, Discounting Victims of Tiger Attacks," *India Forum,* October 30, 2023, https://www.theindiaforum.in/sites/default/files/article_pdf/2023/10/31/1457 -1698724211.pdf.

273   **25 percent surge in the number of sex workers:** R. Mitra, "In the Sundarban, Climate Change Has an Unlikely Effect—on Child Sex-Trafficking," Fuller Project, January 20, 2022.

277   **lost 440 square kilometers:** X. Lu, "The World's Largest Mangrove Forest Is Shrinking," Earth.org, October 8, 2019, https://earth.org/the-worlds-largest -mangrove-forest-is-shrinking/.

278   **$543.3 million in savings:** A. H. M. Raihan Sarker et al., "Value of the Storm-Protection Function of Sundarban Mangroves in Bangladesh," *Journal of Sustainable Development* 13, no. 3 (2020): 128–37.

278   **lost more than $3 billion in value:** B. Bera et al., "Significant Reduction of Carbon Stocks and Changes of Ecosystem Service Valuation of Indian Sundarban," *Scientific Reports* 12, no. 1 (2022): 7809.

278   **Western biologists into nature's saviors:** P. Kareiva and M. Marvier, "What Is Conservation Science?" *BioScience* 62, no. 11 (2012): 962–69.

279   **$125 trillion a year:** R. Costanza et al., "Changes in the Global Value of Ecosystem Services," *Global Environmental Change* 26 (2014): 152–58.

281   **the wild animals city dwellers are most likely to encounter:** U.S. Fish and Wildlife Service et al., *2016 National Survey of Fishing, Hunting, and Wildlife-Associated Recreation* (April 2018), https://www.fws.gov/sites/default/files /documents/news-attached-files/nat_survey2016.pdf.

281   **many studies outlining the ecosystem services:** I'll limit myself to just one

example: In 1990, the introduction of a nonsteroidal painkiller to treat sick cattle in India led to widespread vulture deaths, for whom the medication was poisonous. In all, over 50 million vultures were killed, as their primary source of food was cattle carcasses. Soon after, stray dogs proliferated, spreading rabies, while landfills became hot spots of diseases emanating from the rotting animal flesh that would have otherwise been picked clean. The upshot is that more than 500,000 children died in India in areas where vultures were poisoned and killed, the bulk occurring in the country's most densely packed urban centers.

281   **different species overlapping in a chorus:** M. Hedblom et al., "Bird Diversity Improves the Well-Being of City Residents," in *Ecology and Conservation of Birds in Urban Environments,* ed. E. Murgui and M. Hedblom (Springer, 2017), 287–306.

281   **petrichor:** E. H. Polak and J. Provasi, "Odor Sensitivity to Geosmin Enantiomers," *Chemical Senses* 17, no. 1 (1992): 23–26.

281   **biophilia:** E. O. Wilson, *Biophilia* (Harvard University Press, 1986).

282   **pest control in U.S. cotton fields:** C. Tuneu-Corral et al., "Pest Suppression by Bats and Management Strategies to Favour It: A Global Review," *Biological Reviews* 98, no. 5 (2023): 1564–82.

Epilogue: Humans, a Species of Least Concern

286   **Modern Nature:** D. Jarman, *Modern Nature: Journals, 1989–1990* (Random House, 2017).

Index

about the author

DAN WERB, PhD, is an award-winning writer and social epidemiologist whose work—which primarily investigates the link between big events and human society—has appeared in *The New York Times, Time, The Believer,* and many other outlets. He is an assistant professor in the Division of Infectious Diseases and Global Public Health at the University of California San Diego and in the Dalla Lana School of Public Health at the University of Toronto. Werb is the author of two previous books, *City of Omens* and the award-winning *The Invisible Siege*.

about the type

Birka, named after the historic Viking town, is one of thirty-eight typefaces created by Swedish designer Franko Luin. The font is Luin's first original design, and he credits Garamond as his inspiration. Birka is regarded for its Scandinavian aesthetic.